LEGENDRE POLYNOMIALS
AND FUNCTIONS

by
REFAAT A. EL ATTAR
rea5@hotmail.com
Professor of Mathematics
Faculty of Engineering
Alexandria University
Egypt

Refaat El Attar

ISBN: 1441-49012-4

EAN-13: 978-1441-49012-4

Printed in the United State of America

Copyright © 2009 by the author. All rights reserved. No part of this book may be reproduced, stored in a retrieval system, or transcribed in any form or by any means – electronic, mechanical, photocopying, recording, or otherwise - without the prior permission of the author.

COPYRIGHT © 2009 BY REFAAT EL ATTAR

Preface

This book is written to provide an easy to follow study on the subject of Legendre Polynomials and Functions. It is also written in a way that it can be used as a self study text. Basic knowledge of calculus and differential equations is needed. The book is intended to help students in engineering, physics and applied sciences understand various aspects of Legendre Polynomials and Functions that very often occur in engineering, physics, mathematics and applied sciences.

I have collected many problems and gave numerous solved examples on the subject that might help the reader getting on-hand experience with the techniques presented in this note. It is hoped that this work will give some motivation to the reader to dig a bit further in the subject.

REFAAT A. EL ATTAR
rea5@hotmail.com
Alexandria, EGYPT, February, 2009

Refaat El Attar

Contents

	Page
Legendre Polynomials	7
1. Introduction	9
2. Legendre Differential Equation and Its Solutions	9
3. Legendre Polynomials in Descending Powers of x	17
4. Generating Function for Legendre Polynomials	29
5. Recurrence Relations for Legendre Polynomials	38
6. Orthogonality Properties of Legendre Polynomials	45
7. Integral Form of Legendre Polynomials	55
8. Differential Form for Legendre Polynomials (Rodrigues' Formula)	58
9. Schläfli's Integral for Legendre Polynomials	68
10. Associated Legendre Functions	69
11. Series of Legendre Polynomials	73
12. Legendre Functions of the Second Kind $Q_n(x)$	76
12.1. Relation between $P_n(x)$ and $Q_n(x)$	78
12.2. Properties of Legendre Functions of the Second Kind	79
13. Shifted Legendre Polynomials	82
14. Summary of Legendre Polynomials and Functions	83
Exercises	86
References	92

Refaat El Attar

Legendre Polynomials

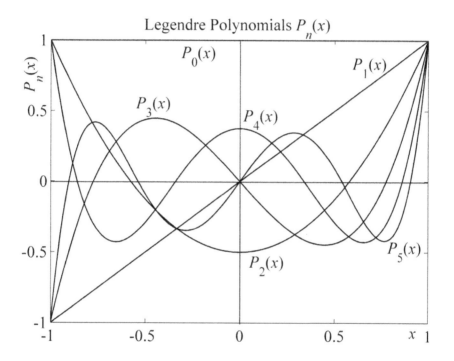

$$P_n(x) = \sum_{k=0}^{N} (-1)^k \cdot \frac{(2n-2k)!}{2^n\, k!\,(n-2k)!\,(n-k)!}\, x^{n-2k}$$

Refaat El Attar

Legendre Polynomials and Functions

1. Introduction

Legendre[1] *Polynomials*, also called *Legendre functions* are solutions of *Legendre differential equation*. They are *Orthogonal Polynomials* and are special case of the *Ultraspherical Functions* and *Jacobi*[2] *Polynomials*. They arise in numerous problems especially in those involving spheres or spherical coordinates or exhibiting spherical symmetry. In spherical polar coordinates, the angular dependence is always best handled by spherical harmonics that are defined in terms of Legendre functions. We start our study with the Legendre differential equation.

2. Legendre Differential Equation and its Solutions

Legendre Differential equation is given by:

$$(1-x^2)y'' - 2xy' + n(n+1)y = 0 \qquad (1)$$

where n is a real number. The solutions of this equation are called *Legendre Functions* of order n. When n is a non-negative integer, the Legendre Functions are often referred to as Legendre Polynomials as we will establish. Since the Legendre differential equation is a second order ordinary differential equation, it has two linearly independent solutions.

It is sometimes useful to write Legendre Differential equation in the form

$$\frac{d}{dx}\left[(1-x^2)\frac{dy}{dx}\right] + n(n+1)y = 0. \qquad (2)$$

It is clear that the only finite singularities of the differential equation are at $x = \pm 1$. In fact $x = \pm 1$ are regular points of the differential equation. Also, it is easy to show that the point at infinity is a regular point.

In order to investigate the solutions of Legendre differential equation, let us first consider the simple case where $n = 0$. In this case, we have

---

1. Adrien-Marie Legendre (1752-1833 Paris-France).
2. Carl Gustav Jacob Jacobi (1804 Potsdam Prussia – 1851 Berlin Germany).

$$\frac{d}{dx}\left[(1-x^2)\frac{dy}{dx}\right] = 0. \tag{3}$$

This is a simple differential equation whose solution is obtained easily. Integrating once, we get

$$(1-x^2)\frac{dy}{dx} = A. \tag{4}$$

Now, by separating the variables, we obtain

$$\int dy = \int \frac{A}{(1-x^2)} dx, \tag{5}$$

giving the solution

$$y = A \cdot \frac{1}{2}\ln\left(\frac{1+x}{1-x}\right) + B. \tag{6}$$

Since this Legendre differential equation is a second order differential equation, its solution contains two arbitrary constants, written here as A and B. And we shall denote the two linearly independent solutions by

$$P_0(x) = 1 \text{ and } Q_0(x) = \frac{1}{2}\ln\left(\frac{1+x}{1-x}\right), \tag{7}$$

where the subscript 0 represents the value of n. It is to be noticed that the second solution $Q_0(x)$ diverges at $x = 1$.

Since the Legendre equation is homogeneous, its general solution is the superposition of $P_0(x)$ and $Q_0(x)$,

$$y = AP_0(x) + BQ_0(x) \tag{8}$$

For the general case where $n \neq 0$, we will use a power series method to solve the Legendre differential equation.

To obtain a series solution about $x = 0$ (an ordinary point), we use the method of undetermined coefficients by assuming a solution of the form:

$$y = \sum_{k=0}^{\infty} c_k x^k \tag{9}$$

Differentiating twice, we obtain

$$y' = \sum_{k=1}^{\infty} k c_k x^{k-1} = \sum_{k=0}^{\infty} (k+1) c_{k+1} x^k \qquad (10)$$

$$y'' = \sum_{k=1}^{\infty} k(k+1) c_{k+1} x^{k-1} = \sum_{k=0}^{\infty} (k+1)(k+2) c_{k+2} x^k \qquad (11)$$

If this assumption is true, then y must satisfy the differential equation. Substituting in the differential equation, we obtain:

$$(1-x^2) \sum_{k=0}^{\infty} (k+1)(k+2) c_{k+2} x^k$$

$$- 2x \sum_{k=0}^{\infty} (k+1) c_{k+1} x^k + n(n+1) \sum_{k=0}^{\infty} c_k x^k = 0$$

The coefficients of each power of x on the left hand side of this equation must be zero, then equating the coefficients of x^k to zero, we obtain

$$(k+1)(k+2) c_{k+2} - k(k-1) c_k - 2k c_k + n(n+1) c_k = 0. \qquad (12)$$

Rearranging, we get

$$\boxed{c_{k+2} = -\frac{(k+n+1)(n-k)}{(k+1)(k+2)} c_k} \qquad (13)$$

This is the *Recurrence Relation for the Coefficients*. Since $x = 0$ is an ordinary point of the differential equation, the two linearly independent solutions can be obtained with the help of this relation. Moreover, if we let c_0 and c_1 be the two arbitrary constants of the solution, then each c_k, $k \geq 2$ can be obtained in terms of either c_0 or c_1.

Now, for even values of k, we have

$$c_2 = -\frac{(n+1)n}{1 \cdot 2} c_0$$

$$c_4 = -\frac{(n+3)(n-2)}{3 \cdot 4} c_2 = \frac{(n+3)(n+1) \cdot n(n-2)}{1 \cdot 2 \cdot 3 \cdot 4} c_0$$

$$c_6 = -\frac{(n+5)(n-4)}{5 \cdot 6} c_4 = -\frac{(n+5)(n+3)(n+1) \cdot n(n-2)(n-4)}{1 \cdot 2 \cdot 3 \cdot 4 \cdot 5 \cdot 6} c_0$$

and so on. In general for the even-suffixed coefficients, we have

$$c_{2k} = (-1)^k \frac{(n+2k-1)(n+2k-3)\cdots(n+1)\cdot n(n-2)\cdots(n-2k+2)}{(2k)!} c_0$$

Similarly, for odd-suffixed coefficients, we obtain

$$c_3 = -\frac{(n+2)(n-1)}{2\cdot 3} c_1$$

$$c_5 = -\frac{(n+4)(n-3)}{4\cdot 5} c_3 = \frac{(n+4)(n+2)\cdot(n-1)(n-3)}{2\cdot 3\cdot 4\cdot 5} c_1$$

$$c_7 = -\frac{(n+6)(n-5)}{6\cdot 7} c_5 = -\frac{(n+6)(n+4)(n+2)\cdot(n-1)(n-3)(n-5)}{2\cdot 3\cdot 4\cdot 5\cdot 6\cdot 7} c_1$$

Then,

$$c_{2k+1} = (-1)^k \frac{(n+2k)(n+2k-2)\cdots(n+2)\cdot(n-1)(n-3)\cdots(n-2k+1)}{(2k+1)!} c_1$$

The solutions can now be obtained by making choices for c_0 and c_1. If $c_0 = 1$ and $c_1 = 0$, the first solution will be

$$y_1(x) = \sum_{k=0}^{\infty} (-1)^k \frac{(n+2k-1)(n+2k-3)\cdots(n+1)\cdot n(n-2)\cdots(n-2k+2)}{(2k)!} x^{2k}$$

(14)

On the other hand, If $c_0 = 0$ and $c_1 = 1$, the second solution will be

$$y_2(x) = \sum_{k=0}^{\infty} (-1)^k \frac{(n+2k)(n+2k-2)\cdots(n+2)\cdot(n-1)(n-3)\cdots(n-2k+1)}{(2k+1)!} x^{2k+1}$$

(15)

And since the nearest singularities are at $x = \pm 1$, then these two series solutions converge for all $x \in (-1, 1)$. Moreover, since the first solution is in even powers of x, while the second is in odd powers of x, the two solutions are in fact linearly independent.

The general solution of Legendre differential equation is given as a linear combination of the two linearly independent series solutions. Writing the first few terms in the general solution

$$y(x) = A\left\{1 - \frac{n\cdot(n+1)}{2!}x^2 + \frac{n(n-2)\cdot(n+1)(n+3)}{4!}x^4 - \cdots\right\}$$
$$+ B\left\{x - \frac{(n-1)\cdot(n+2)}{3!}x^3 + \frac{(n-1)(n-3)\cdot(n+2)(n+4)}{5!}x^5 - \cdots\right\} \quad (16)$$

Now, looking closely at each of the two solutions, we can observe that if n is an even integer, the first series reduces to an even polynomial of degree n. For example, if $n = 2$, the first solution, apart from the multiplicative constant, becomes

$$y_1(x) = 1 - 3x^2$$

While, if n is an odd integer, the second series reduces to an odd polynomial of degree n. For example, if $n = 3$, the second solution, apart from the multiplicative constant, becomes

$$y_2(x) = x - \frac{5}{3}x^3$$

These polynomial solutions, apart from the arbitrary constants, are called *Legendre Polynomials* or *Legendre Functions of the First Kind*. They are denoted by $P_n(x)$. We can say that $P_n(x)$ is the first solution of Legendre Differential Equation. For the second solution, if n is an even integer, $y_2(x)$ remains an infinite series, while if n is an odd integer, $y_1(x)$ is again an infinite series. In this case, these infinite series represent the second solution for Legendre Differential equation. They are denoted by $Q_n(x)$, and are called *Legendre Functions of the Second Kind*. We will get to them later. The general solution can now be written as

$$\boxed{y(x) = aP_n(x) + bQ_n(x)} \quad (17)$$

The second solution $Q_n(x)$ can be expressed in terms of the first solution $P_n(x)$ from the analysis of ordinary differential equations as

$$Q_n(x) = P_n(x)\int \frac{1}{(1-x^2)[P_n(x)]^2}dx \quad . \quad (18)$$

We will study this later.

Example 1: Find the general solution of the Legendre equation

$$(1-x^2)y'' - 2xy' + 2y = 0.$$

Solution: Comparing this equation with Legendre Differential Equation

$(1-x^2)y'' - 2xy' + n(n+1)y = 0$, we have $n(n+1) = 2$,

then $n = 1$, and the first solution is

$y_1 = P_1(x) = x$.

For the second solution, we have

$$y_2 = P_1(x)\int \frac{dx}{(1-x^2)[P_1(x)]^2} = x\int \frac{dx}{(1-x^2)x^2}$$

$$= \frac{x}{2}\int \left(\frac{1}{1+x} + \frac{1}{1-x} + \frac{2}{x^2}\right)dx = \left(\frac{x}{2}\ln\left[\frac{1+x}{1-x}\right] - 1\right).$$

And the general solution will be $y = ax + b\left(\frac{x}{2}\ln\left[\frac{1+x}{1-x}\right] - 1\right)$. □

Example 2: Obtain one series solution for Legendre Differential Equation

$$(1-x^2)y'' - 2xy' + n(n+1)y = 0 \text{ about } x = 1.$$

Solution: $x = 1$ is a regular point of the differential equation, then we will use the method of Frobenius to obtain a series solution. But first, let us make the substitution $t = x - 1$. This will simplify somehow the procedure. Now,

$$\frac{dy}{dx} = \frac{dy}{dt}\cdot\frac{dt}{dx} = \frac{dy}{dt},$$

and $\frac{d^2y}{dx^2} = \frac{d}{dx}\left(\frac{dy}{dt}\right) = \frac{d}{dt}\left(\frac{dy}{dt}\right)\cdot\frac{dx}{dt} = \frac{d^2y}{dt^2}.$

Substituting in the differential equation, we get

$$[1-(1+t)^2]\frac{d^2y}{dt^2} - 2(1+t)\frac{dy}{dt} + n(n+1)y = 0,$$

or

Legendre Polynomials and Functions

$$t(t+2)\frac{d^2y}{dt^2} + 2(1+t)\frac{dy}{dt} - n(n+1)y = 0.$$

$t = 0$ is now a regular point of this differential equation, and we can assume a series solution of the form $y = \sum_{k=0}^{\infty} c_k t^{k+\alpha}$.

Differentiating twice, we obtain

$$\frac{dy}{dt} = \sum_{k=0}^{\infty} (k+\alpha) c_k t^{k+\alpha-1}, \text{ and}$$

$$\frac{d^2y}{dt^2} = \sum_{k=0}^{\infty} (k+\alpha-1)(k+\alpha) c_k t^{k+\alpha-2}.$$

Substituting all these expressions in to differential equation, we obtain

$$t(t+2) \sum_{k=0}^{\infty} (k+\alpha-1)(k+\alpha) c_k t^{k+\alpha-2}$$

$$+ 2(t+1) \sum_{k=0}^{\infty} (k+\alpha) c_k t^{k+\alpha-1} + n(n+1) \sum_{k=0}^{\infty} c_k t^{k+\alpha} = 0$$

Equating the coefficients of the least powr of t ($t^{\alpha-1}$) to zero, we get $2\alpha(\alpha-1)c_0 + 2\alpha c_0 = 0$, and since $c_0 \neq 0$, we have $\boxed{\alpha = 0, 0}$ (a double root). Then Frobenius method will produce one solution.

Equating the coefficients of $t^{k+\alpha}$ to zero, we obtain
$(k+\alpha-1)(k+\alpha)c_k + 2(k+\alpha)(k+\alpha+1)c_{k+1}$

$$+ 2(k+\alpha)c_k + 2(k+\alpha+1)c_{k+1} - n(n+1)c_k = 0.$$

Re-arranging, we get the recurrence relation for the coefficients as

$$\boxed{c_{k+1} = -\frac{(k+\alpha)(k+\alpha+1) - n(n+1)}{2(k+\alpha+1)^2} c_k}.$$

For $\alpha = 0$, this recurrence relation becomes

$$c_{k+1} = -\frac{k(k+1) - n(n+1)}{2(k+1)^2} c_k$$

For various values of k, we have

$$k = 0: \quad c_1 = \frac{n(n+1)}{2} c_0$$

$$k = 1: \quad c_2 = \frac{1 \cdot 2 - n(n+1)}{2 \cdot 2^2} c_1 = \frac{n(n-1) \cdot (n+1)(n+2)}{2^2 \cdot (1 \cdot 2)^2} c_0$$

$$k = 2: \quad c_3 = \cdots = \frac{n(n-1)(n-2) \cdot (n+1)(n+2)(n+3)}{2^3 \cdot (1 \cdot 2 \cdot 3)^2} c_0.$$

And in general

$$c_k = \frac{n(n-1)\cdots(n-k+1) \cdot (n+1)(n+2)\cdots(n+k)}{2^k \cdot (k!)^2} c_0.$$

Using the Pochhammer[3] symbol, we can write

$$c_k = \frac{(n)_k \cdot (n+1)_k}{2^k \cdot (k!)^2} c_0, \text{ and the series solution is now}$$

$$y = c_0 \left[1 + \sum_{k=1}^{\infty} \frac{(n)_k \cdot (n+1)_k}{2^k \cdot (k!)^2} t^k \right].$$

Back-substituting, we obtain

$$y = c_0 \left[1 + \sum_{k=1}^{\infty} \frac{(n)_k \cdot (n+1)_k}{(k!)^2} \cdot \left(\frac{x-1}{2} \right)^k \right]. \qquad \square$$

---

3. Leo August Pochhammer (1841-1920). The Pochhammer symbol is given by
$$(n)_k = \frac{\Gamma(n+k)}{\Gamma(n)} = n(n+1)(n+2)\cdots(n+k-1), \quad (n)_0 = 1$$

3. Legendre Polynomials in Descending Powers of x

It is sometimes advantageous to write the polynomial solutions in descending powers of x. From the previous section, rewriting the recurrence relation for the coefficients as

$$c_k = -\frac{(k+2)(k+1)}{(k+n+1)(n-k)} c_{k+2}, \quad k = 0, 1, 2, \cdots \tag{19}$$

If $k = n$, then we have $c_{n+2} = c_{n+4} = c_{n+6} = \cdots = 0$.

Therefore for $k = n-2, n-4, n-6, \cdots$, we find that

$$c_{n-2} = -\frac{n(n-1)}{2(2n-1)} c_n$$

$$c_{n-4} = -\frac{(n-2)(n-3)}{4(2n-3)} c_{n-2} = \frac{n(n-1)(n-2)(n-3)}{2 \cdot 4 \cdot (2n-1)(2n-3)} c_n$$

$$c_{n-6} = \cdots = -\frac{n(n-1)(n-2)(n-3)(n-4)(n-5)}{2 \cdot 4 \cdot 6 \cdot (2n-1)(2n-3)(2n-5)} c_n$$

and so on. Then, the polynomial takes the form

$$y(x) = c_n \left\{ \begin{array}{l} x^n - \dfrac{n(n-1)}{2(2n-1)} x^{n-2} + \dfrac{n(n-1)(n-2)(n-3)}{2 \cdot 4 \cdot (2n-1)(2n-3)} x^{n-4} \\ \quad - \dfrac{n(n-1)(n-2)(n-3)(n-4)(n-5)}{2 \cdot 4 \cdot 6 \cdot (2n-1)(2n-3)(2n-5)} x^{n-6} + \cdots \end{array} \right\}$$

$$(20)$$

To normalize these polynomials, we chose the arbitrary constant c_n, the coefficient of x^n, so that $P_n(1) = 1$. The value of c_n to achieve this is

$$c_n = \frac{(2n-1)(2n-3) \cdots 3 \cdot 1}{n!} \tag{21}$$

Then, the Legendre Polynomials are

$$P_n(x) = \frac{(2n-1)(2n-3) \cdots 3 \cdot 1}{n!} \left\{ x^n - \frac{n(n-1)}{2(2n-1)} x^{n-2} + \frac{n(n-1)(n-2)(n-3)}{2 \cdot 4 \cdot (2n-1)(2n-3)} x^{n-4} - \cdots \right\}$$

To find a compact form for $P_n(x)$, the general term in this last series is

$$(-1)^k \frac{1 \cdot 3 \cdot 5 \cdots (2n-1)}{n!} \cdot \frac{n(n-1)(n-2) \cdots (n-2k+1)}{2 \cdot 4 \cdot 6 \cdots 2k \cdot (2n-1)(2n-3) \cdots (2n-2k+1)} x^{n-2k}$$

Now, we do some tricks to make this general term more compact. We have

$$1 \cdot 3 \cdot 5 \cdots (2n-1) = \frac{1 \cdot 2 \cdot 3 \cdot 4 \cdots (2n-1) \cdot 2n}{2 \cdot 4 \cdot 6 \cdots 2n} = \frac{(2n)!}{2^n n!}$$

$n(n-1)(n-2)\cdots(n-2k+1)$

$$= \frac{n(n-1)(n-2)\cdots(n-2k+1)(n-2k)(n-2k-1)\cdots 3 \cdot 2 \cdot 1}{(n-2k)(n-2k-1)\cdots 3 \cdot 2 \cdot 1} = \frac{n!}{(n-2k)!}$$

$(2n-1)(2n-3)\cdots(2n-2k+1)$

$$= \frac{2n(2n-1)(2n-2)\cdots(2n-2k+2)(2n-2k+1)}{2n(2n-2)(2n-4)\cdots(2n-2k+2)} \times \frac{(2n-2k)!}{(2n-2k)!}$$

$$= \frac{(2n)!}{2^n n(n-1)(n-2)\cdots(n-k+1)(2n-2k)!}$$

$$= \frac{(2n)!}{2^n (2n-2k)!} \cdot \frac{(n-k)(n-k-1)\cdots 3 \cdot 2 \cdot 1}{n(n-1)(n-2)\cdots(n-k+1)(n-k-1)\cdots 3 \cdot 2 \cdot 1}$$

$$= \frac{(2n)!(n-k)!}{2^n (2n-2k)! n!}$$

Substituting all these values, the general term becomes

$$(-1)^k \cdot \frac{(2n-2k)!}{2^n k!(n-2k)!(n-k)!} x^{n-2k}$$

And since the polynomial is of degree n, k must be chosen so that $n-2k \geq 0$, i.e., $k \leq n/2$. Then, if n is even, k goes from 0 to $n/2$, while if n is odd, k goes from 0 to $(n-1)/2$. Hence $P_n(x)$ can now be written as

$$\boxed{P_n(x) = \sum_{k=0}^{N} (-1)^k \cdot \frac{(2n-2k)!}{2^n k!(n-2k)!(n-k)!} x^{n-2k}} \qquad (22)$$

where N is called the *Floor Function* and is given by

$$N = \begin{cases} n/2 & \text{if } n \text{ is even} \\ (n-1)/2 & \text{if } n \text{ is odd} \end{cases}$$

Here are few of these polynomials for different values of n:

$P_0(x) = 1$;

$P_1(x) = x$;

$$P_2(x) = \frac{1}{2}(3x^2 - 1);$$

$$P_3(x) = \frac{1}{2}(5x^3 - 3x);$$

$$P_4(x) = \frac{1}{8}(35x^4 - 30x^2 + 3);$$

$$P_5(x) = \frac{1}{8}(63x^5 - 70x^3 + 15x);$$

$$P_6(x) = \frac{1}{16}(231x^6 - 315x^4 + 105x^2 - 5).$$

Note: Although Legendre Polynomials are defined for all finite values of x, they are solutions of Legendre differential equation only for $x \in (-1, 1)$.

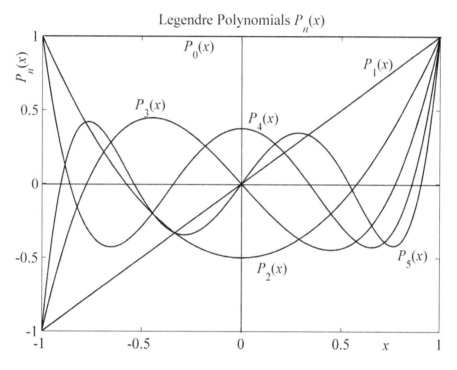

Legendre Polynomials $P_n(x)$

Legendre Polynomials can be generated using the *Gram-Schmidt*[4] *Orthogonalization Process* as follows: Given a set of linearly independent

---

4. Jørgen Pedersen Gram (1850–1916 Denmark), Erhard Schmidt (1876-1959 Germany)

functions $\{1, x, x^2, x^3, \cdots\}$, we can construct Legendre Polynomials in the interval $(-1, 1)$. The steps are

$$P_0(x) = 1$$

$$P_1(x) = \left(x - \frac{\int_{-1}^{1} x P_0^2(x)\,dx}{\int_{-1}^{1} P_0^2(x)\,dx} \right) P_0(x) = x$$

$$P_k(x) = \left(x - \frac{\int_{-1}^{1} x P_{k-1}^2(x)\,dx}{\int_{-1}^{1} P_{k-1}^2(x)\,dx} \right) P_{k-1}(x) - \left(\frac{\int_{-1}^{1} P_{k-1}^2(x)\,dx}{\int_{-1}^{1} P_{k-2}^2(x)\,dx} \right) P_{k-2}(x),$$

$$k = 2, 3, 4, \cdots$$

Example 3: Express the function $f(x) = 5x^3 + 6x^2 + 7x + 2$ in terms of Legendre Polynomials.

Solution: We have $P_0(x) = 1$; $P_1(x) = x$; $P_2(x) = \frac{1}{2}(3x^2 - 1)$ and

$P_3(x) = \frac{1}{2}(5x^3 - 3x)$. Then

$$f(x) = 5x^3 + 6x^2 + 7x + 2 = \frac{a_3}{2}(5x^3 - 3x) + \frac{a_2}{2}(3x^2 - 1) + a_1 x + a_0$$

Equating the coefficients of various powers of x in both sides, we obtain

coef. of x^3: $5 = \frac{5}{2} a_3$ $a_3 = 2$

coef. of x^2: $6 = \frac{3}{2} a_2$ $a_2 = 4$

coef. of x^1: $7 = -3 + a_1$ $a_1 = 10$

coef. of x^0: $2 = -2 + a_0$ $a_0 = 4$

Substituting all these values, we get

$$f(x) = 2P_3(x) + 4P_2(x) + 10P_1(x) + 4P_0(x). \qquad \square$$

Legendre Polynomials and Functions

Example 4: Show that: $\sin^4\theta = \dfrac{8}{15}P_0(\cos\theta) + \dfrac{16}{21}P_2(\cos\theta) + \dfrac{8}{35}P_4(\cos\theta)$.

Solution: We have

$$\sin^4\theta = (1-\cos^2\theta)^2 = 1 - 2\cos^2\theta + \cos^4\theta$$

$$= a\, P_0(\cos\theta) + b\, P_2(\cos\theta) + c\, P_4(\cos\theta)$$

Also,

$$P_0(\cos\theta) = 1;\quad P_2(\cos\theta) = \dfrac{1}{2}(3\cos^2\theta - 1)$$

$$P_4(\cos\theta) = \dfrac{1}{8}(35\cos^4\theta - 30\cos^2\theta + 3).$$

Therefore,

$$\sin^4\theta = 1 - 2\cos^2\theta + \cos^4\theta$$

$$= a + \dfrac{b}{2}(3\cos^2\theta - 1) + \dfrac{c}{8}(35\cos^4\theta - 30\cos^2\theta + 3)$$

Equating various coefficients, we get

$$a = \dfrac{8}{15},\ b = \dfrac{16}{21},\ c = \dfrac{8}{35}.$$

Therefore,

$$\sin^4\theta = \dfrac{8}{15}P_0(\cos\theta) + \dfrac{16}{21}P_2(\cos\theta) + \dfrac{8}{35}P_4(\cos\theta).\qquad\square$$

In some applications, it is more convenient to use $P_n(\cos\theta)$. The graph is given below.

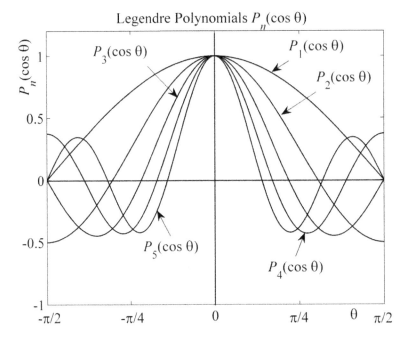

Example 5: In Legendre differential equation, use the substitution $x = \cos\theta$ to show that:

i. $\dfrac{d^2 y}{d\theta^2} + \cot\theta \dfrac{dy}{d\theta} + n(n+1)y = 0$.

ii. $\dfrac{1}{\sin\theta} \cdot \dfrac{d}{d\theta}\left(\sin\theta \dfrac{dy}{d\theta}\right) + n(n+1)y = 0$.

Solution: i. We have: $\dfrac{dy}{dx} = \dfrac{dy}{d\theta} \cdot \dfrac{d\theta}{dx} = -\operatorname{cosec}\theta \cdot \dfrac{dy}{d\theta}$, and

$$\dfrac{d^2 y}{dx^2} = \dfrac{d}{dx}\left(\dfrac{dy}{dx}\right) = \dfrac{d}{dx}\left(-\operatorname{cosec}\theta \cdot \dfrac{dy}{d\theta}\right) = \dfrac{d}{d\theta}\left(-\operatorname{cosec}\theta \cdot \dfrac{dy}{d\theta}\right) \cdot \dfrac{d\theta}{dx}$$

$$= \operatorname{cosec}^2\theta \cdot \dfrac{d^2 y}{d\theta^2} - \operatorname{cosec}^2\theta \cot\theta \cdot \dfrac{dy}{d\theta}.$$

Now, from Legendre differential equation

$$(1-x^2)\frac{d^2y}{dx^2} - 2x\frac{dy}{dx} + n(n+1)y = 0,$$

we substitute all these values to obtain

$$\sin^2\theta\left(\operatorname{cosec}^2\theta\cdot\frac{d^2y}{d\theta^2} - \operatorname{cosec}^2\theta\cot\theta\cdot\frac{dy}{d\theta}\right)$$

$$-\cos\theta\left(-\operatorname{cosec}\theta\cdot\frac{dy}{d\theta}\right) + n(n+1)y = 0$$

Rearranging, we get $\dfrac{d^2y}{d\theta^2} + \cot\theta\dfrac{dy}{d\theta} + n(n+1)y = 0$.

ii. We have

$$\frac{1}{\sin\theta}\frac{d}{d\theta}\left(\sin\theta\,\frac{dy}{d\theta}\right) + n(n+1)y$$

$$= \frac{1}{\sin\theta}\left(\sin\theta\,\frac{d^2y}{d\theta^2} + \cos\theta\,\frac{dy}{d\theta}\right) + n(n+1)y$$

$$= \frac{d^2y}{d\theta} + \cot\theta\,\frac{dy}{d\theta} + n(n+1)y = 0. \qquad \square$$

Example 6: Show that all the roots of $P_n(x) = 0$ are distinct.

Solution: We will show this by contradiction. Suppose that we have a double root at $x = \alpha$, then from the theory of equations, this root must satisfy the two equations: $P_n(\alpha) = 0$ and $P_n'(\alpha) = 0$.

Since $P_n(x)$ satisfies Legendre differential equation, we have

$$(1-x^2)P_n''(x) - 2xP_n'(x) + n(n+1)P_n(x) = 0.$$

Differentiating m times and using Leibniz Theorem[5], we get

---

5. **Leibniz Theorem:** If $u(x)$ and $v(x)$ are continuously differentiable, then the mth derivative of their product is given by:

$$(u\cdot v)_m = u\cdot v_m + m\cdot u_1\cdot v_{m-1} + \frac{m(m-1)}{2!}\cdot u_2\cdot v_{m-2} + \cdots + u_m\cdot v,$$

where u_m and v_m are the mth derivatives of $u(x)$ and $v(x)$.

$$(1-x^2)\frac{d^{m+2}}{dx^{m+2}}P_n(x) - 2x(m+1)\frac{d^{m+1}}{dx^{m+1}}P_n(x)$$

$$+[n(n+1)-m(m+1)]\frac{d^m}{dx^m}P_n(x) = 0$$

Letting $m=0$ and $x=\alpha$ in this last equation, we get

$(1-\alpha^2)P_n''(\alpha) - 2\alpha P_n'(\alpha) + n(n+1)P_n(\alpha) = 0$.

But, $P_n(\alpha) = 0$ and $P_n'(\alpha) = 0$,

then we have, $(1-\alpha^2)P_n''(\alpha) = 0$ or $P_n''(\alpha) = 0$.

Similarly, Let $m=1$ and $x=\alpha$, we conclude that $P_n'''(\alpha) = 0$.

If we continue with this process, we obtain for $m = n-2$:

$$\left.\frac{d^n}{dx^n}P_n(x)\right|_{x=\alpha} = 0.$$

But $P_n(x) = \dfrac{(2n-1)(2n-3)\cdots 3\cdot 1}{n!}\cdot\left\{x^n - \dfrac{n(n-1)}{2(2n-1)}x^{n-2} + \cdots\right\}$.

Then $\left.\dfrac{d^n}{dx^n}P_n(x)\right|_{x=\alpha} = \dfrac{(2n-1)(2n-3)\cdots 3\cdot 1}{n!}\cdot n!$.

This is a contradiction. Then our assumption is not true, and all the roots of $P_n(x) = 0$ must be distinct. □

Example 7: Show that $\displaystyle\int_x^1 P_n(x)\,dx = \dfrac{(1-x^2)P_n'(x)}{n(n+1)}$.

Solution: Since $P_n(x)$ satisfies Legendre Differential Equation, then

$$P_n = -\frac{1}{n(n+1)}\left\{(1-x^2)P_n'' - 2x\,P_n'\right\}.$$

Integrating with respect to x from x to 1, we get

$$I = \int_x^1 P_n\,dx = -\frac{1}{n(n+1)}\int_x^1\left\{(1-x^2)P_n'' - 2x\,P_n'\right\}dx$$

$$= -\frac{1}{n(n+1)} \left\{ \int_x^1 (1-x^2) P_n'' \, dx - 2\int_x^1 x \, P_n' \, dx \right\}$$

For the first integral, using integration by parts, we get

$$I = -\frac{1}{n(n+1)} \left\{ \int_x^1 (1-x^2) dP_n' - 2\int_x^1 x \, P_n' \, dx \right\}$$

$$= -\frac{1}{n(n+1)} \left\{ \left[(1-x^2) P_n'\right]_x^1 - \int_x^1 P_n' \, d(1-x^2) - 2\int_x^1 x \, P_n' \, dx \right\}$$

$$= -\frac{1}{n(n+1)} \left\{ -(1-x^2) P_n' + 2\int_x^1 x \, P_n' \, dx - 2\int_x^1 x \, P_n' \, dx \right\}$$

$$= \frac{(1-x^2) P_n'}{n(n+1)}.$$

Therefore, $\int_x^1 P_n(x) \, dx = \dfrac{(1-x^2) P_n'(x)}{n(n+1)}$. $\square$

Example 8: Solve Legendre differential equation in descending powers of x:
$$(1-x^2) y'' - 2xy' + n(n+1) y = 0.$$

Solution: Assume a solution of the form $y = \sum_{k=0}^{\infty} c_k x^{\alpha-k}$, then

$$y' = \sum_{k=0}^{\infty} (\alpha - k) c_k x^{\alpha - k - 1}$$

and $y'' = \sum_{k=0}^{\infty} (\alpha - k)(\alpha - k - 1) c_k x^{\alpha - k - 2}$

Substituting in the differential equation, we obtain

$$(1-x^2)\sum_{k=0}^{\infty}(\alpha-k)(\alpha-k-1)c_k x^{\alpha-k-2}$$

$$-2x\sum_{k=0}^{\infty}(\alpha-k)c_k x^{\alpha-k-1}+n(n+1)\sum_{k=0}^{\infty}c_k x^{\alpha-k}=0$$

Equating the coefficients of the *highest power of* x, (x^{α}) to zero, we get

$-\alpha(\alpha-1)c_0-2\alpha c_0+n(n+1)c_0=0$, $c_0\neq 0$, then

$(\alpha-n)(\alpha+n+1)=0$, giving $\boxed{\alpha=n}$ and $\boxed{\alpha=-(n+1)}$.

Equating the coefficients of the *next highest power of* x, ($x^{\alpha-1}$) to zero, we get $-(\alpha-1)(\alpha-2)c_1-2(\alpha-1)c_1+n(n+1)c_1=0$, or

$c_1(n-\alpha+1)(n+\alpha)=0$, but $(n-\alpha+1)(n+\alpha)=0$, then $\boxed{c_1=0}$.

Now, equating the coefficients of the general term $x^{\alpha-k}$ to zero, we get

$(\alpha-k+2)(\alpha-k+1)c_{k-2}$
$\quad -(\alpha-k)(\alpha-k-1)c_k-2(\alpha-k)c_k+n(n+1)c_k=0$

Rearranging, we get

$$\boxed{c_k=\frac{(\alpha-k+2)(\alpha-k+1)-n(n+1)}{(\alpha-k+2)(\alpha-k+1)}c_{k-2}}$$

Since $c_1=0$, the it follows that $c_3=c_5=\cdots=0$

For $\boxed{\alpha=n}$, we have

$$c_k=\frac{(n-k+2)(n-k+1)-n(n+1)}{(n-k+2)(n-k+1)}c_{k-2}$$

$$=-\frac{(n-k+2)(n-k+1)}{k(2n-k+1)}c_{k-2}$$

And the first solution is

$$y_1(x) = c_0 \left\{ x^n - \frac{n(n-1)}{2(2n-1)} x^{n-2} + \frac{n(n-1)(n-2)(n-3)}{2 \cdot 4 \cdot (2n-1)(2n-3)} x^{n-4} \right.$$
$$\left. - \frac{n(n-1)(n-2)(n-3)(n-4)(n-5)}{2 \cdot 4 \cdot 6 \cdot (2n-1)(2n-3)(2n-5)} x^{n-6} + \cdots \right\}$$

If we take, $c_0 = \dfrac{1 \cdot 3 \cdot 5 \cdots (2n-1)}{n!}$, This solution is in fact Legendre Polynomial $P_n(x)$, therefore

$$P_n(x) = \frac{1 \cdot 3 \cdot 5 \cdots (2n-1)}{n!} \left\{ x^n - \frac{n(n-1)}{2(2n-1)} x^{n-2} \right.$$
$$+ \frac{n(n-1)(n-2)(n-3)}{2 \cdot 4 \cdot (2n-1)(2n-3)} x^{n-4}$$
$$\left. - \frac{n(n-1)(n-2) \cdots (n-5)}{2 \cdot 4 \cdot 6 \cdot (2n-1)(2n-3)(2n-5)} x^{n-6} + \cdots \right\}$$

For $\boxed{\alpha = -(n+1)}$, we have

$$c_k = \frac{(n+k-1)(n+k)}{k(2n+k+1)} c_{k-2}$$

And the second solution is

$$y_2(x) = c_0 \left\{ x^{-n-1} + \frac{(n+1)(n+2)}{2(2n+3)} x^{-n-3} \right.$$
$$\left. + \frac{(n+1)(n+2)(n+3)(n+4)}{2 \cdot 4 \cdot (2n+3)(2n+5)} x^{-n-5} + \cdots \right\}$$

If we take, $c_0 = \dfrac{n!}{1 \cdot 3 \cdot 5 \cdots (2n+1)}$, This solution is in fact Legendre Function of the second kind $Q_n(x)$, therefore

$$Q_n(x) = \frac{n!}{1 \cdot 3 \cdot 5 \cdots (2n+1)} \left\{ x^{-n-1} + \frac{(n+1)(n+2)}{2(2n+3)} x^{-n-3} \right.$$
$$\left. + \frac{(n+1)(n+2)(n+3)(n+4)}{2 \cdot 4 \cdot (2n+3)(2n+5)} x^{-n-5} + \cdots \right\}$$

The general solution is then $y(x) = A P_n(x) + B Q_n(x)$. □

Note: From the previous example, Legendre Polynomials as well as Legendre Functions of the second kind can be put in a summation form as

$$\boxed{P_n(x) = \sum_{k=0}^{N} \frac{(-1)^k (2n-2k)! x^{n-2k}}{2^n k! (n-k)!(n-2k)!}},$$

where N is the *Floor Function*

$$N = \begin{cases} n/2 & \text{if } n \text{ is even} \\ (n-1)/2 & \text{if } n \text{ is odd.} \end{cases}$$

This is exactly the expression for $P_n(x)$ as obtain by Equation (22).

For $Q_n(x)$, we have

$$\boxed{Q_n(x) = \frac{2^n n!}{(2n+1)!} \sum_{k=0}^{\infty} \frac{(n+2k)! \cdot x^{-(n+2k+1)}}{2^k k! (2n+3)(2n+5) \cdots (2n+2k+1)}} \qquad (23)$$

4. Generating Function for Legendre Polynomials

Generating functions are available for most special functions and polynomials. We start with a definition.

Definition: Let $f(x,t)$ be a function of x and t that can be expressed in a Taylor's series in t as

$$f(t,x) = \sum_{n=0}^{\infty} t^n C_n(x),$$

then the function $f(x,t)$ is called a **generating function** of the functions $C_n(x)$.

For example, the function

$$f(x,t) = \frac{1}{1-xt} \qquad (24)$$

is a generating function of the polynomials x^n since

$$\frac{1}{1-xt} = \sum_{n=0}^{\infty} t^n x^n, \quad |xt| < 1. \qquad (25)$$

Now, the function

$$g(x,t) = (1 - 2xt + t^2)^{-1/2}, \text{ with } |x| \leq 1, \; |t| < 1 \qquad (26)$$

is the generating function for Legendre polynomials. In fact, we have

$$\boxed{(1 - 2xt + t^2)^{-1/2} = \sum_{n=0}^{\infty} t^n P_n(x)} \qquad (27)$$

To prove this relation, we expand the generating function to the left using the **Binomial Expansion**[6], to obtain

---

6. *Binomial Expansion*:

$$(1+x)^m = 1 + mx + \frac{m(m-1)}{2!}x^2 + \frac{m(m-1)(m-2)}{3!}x^3 + \frac{m(m-1)(m-2)(m-3)}{4!}x^4 + \cdots$$

$$(1-2xt+t^2)^{-1/2} = [1+t(2x-t)]^{-1/2}$$

$$= 1 + \frac{1}{2}t(2x-t) + \frac{1 \cdot 3}{2 \cdot 4}t^2(2x-t)^2 + \cdots + \frac{1 \cdot 3 \cdot 5 \cdots (2n-1)}{2 \cdot 4 \cdot 6 \cdots 2n}t^n(2x-t)^n + \cdots$$

The coefficient of t^n in this expression is

$$\frac{1 \cdot 3 \cdot 5 \cdots (2n-1)}{2 \cdot 4 \cdot 6 \cdots 2n}(2x)^n + \frac{1 \cdot 3 \cdot 5 \cdots (2n-3)}{2 \cdot 4 \cdot 6 \cdots (2n-2)}(n-1)(2x)^{n-2} + \cdots$$

$$= \frac{1 \cdot 3 \cdot 5 \cdots (2n-1)}{n!}x^n + \frac{1 \cdot 3 \cdot 5 \cdots (2n-1)}{n!} \cdot \frac{n(n-1)}{2(2n-1)}x^{n-2} + \cdots$$

$$= \frac{1 \cdot 3 \cdot 5 \cdots (2n-1)}{n!}\left(x^n + \frac{n(n-1)}{2(2n-1)}x^{n-2} + \frac{n(n-1)(n-2)(n-3)}{2 \cdot 4 \cdot (2n-1)(2n-3)}x^{n-4} + \cdots\right)$$

$$= \sum_{k=0}^{N}(-1)^k \cdot \frac{(2n-2k)!}{2^k k!(n-2k)!(n-k)!}x^{n-2k} = P_n(x).$$

where N is the **Floor Function**

$$N = \begin{cases} n/2 & \text{if } n \text{ is even} \\ (n-1)/2 & \text{if } n \text{ is odd} \end{cases} \tag{28}$$

Using this generating function, we can derive many properties of Legendre Polynomials. But first, let us give some illustrative examples to show the use of this generating function.

Example 9: Using the generating function, $(1-2xt+t^2)^{-1/2} = \sum_{n=0}^{\infty} t^n P_n(x)$, or otherwise, show that:

(1) $P_n(1) = 1$ (2) $P_n(-1) = (-1)^n$

(3) $P_n'(1) = \frac{n(n+1)}{2}$ (4) $P_n'(-1) = (-1)^{n+1}\frac{n(n+1)}{2}$

(5) $P_{2n}(0) = (-1)^n \frac{(2n)!}{2^{2n}(n!)^2}$

(6) $P_{2n+1}(0) = 0$ (7) $P_n(-x) = (-1)^n P_n(x)$

Solution: We will start from the generating function:

$$(1+2xt+t^2)^{-1/2} = \sum_{n=0}^{\infty} t^n P_n(x).$$

(1) Letting $x = 1$, we obtain

$$(1-2t+t^2)^{-1/2} = (1-t)^{-1} = \sum_{n=0}^{\infty} t^n P_n(1).$$

Using the Binomial expansion of $(1-t)^{-1}$, we obtain

$$1+t+t^2 + \cdots + t^n + \cdots = \sum_{n=0}^{\infty} t^n P_n(1).$$

Equating the coefficients of t^n in both sides, we get $P_n(1) = 1$. □

(2) Letting $x = -1$ in the generating function, we obtain

$$(1+2t+t^2)^{-1/2} = (1+t)^{-1} = \sum_{n=0}^{\infty} t^n P_n(-1).$$

Using the Binomial expansion of $(1+t)^{-1}$, we obtain

$$1-t+t^2-\cdots+(-1)^n t^n + \cdots = \sum_{n=0}^{\infty} t^n P_n(-1).$$

Equating the coefficients of t^n in both sides, we get

$P_n(-1) = (-1)^n$. □

(3) $P_n(x)$ satisfies the Legendre Differential equation, then

$$(1-x^2)P_n''(x) - 2x P_n'(x) + n(n+1)P_n(x) = 0.$$

Letting $x = 1$, we obtain $2P_n'(1) = n(n+1)P_n(1)$,

and since $P_n(1) = 1$, then $P_n'(1) = \dfrac{n(n+1)}{2}$. □

(4) Letting $x = -1$ in the differential equation, we obtain

$$2P_n'(-1) = -n(n+1)P_n(-1),$$

and since $P_n(-1) = (-1)^n$, then

$$P_n'(-1) = (-1)^{n+1}\dfrac{n(n+1)}{2}.$$ □

(5) Letting $x = 0$ in the generating function, we obtain

$$(1+t^2)^{-1/2} = \sum_{n=0}^{\infty} t^n P_n(0)$$

Using the Binomial expansion of $(1+t^2)^{-1/2}$, we obtain

$$1 - \frac{1}{2} \cdot t^2 + \frac{1\cdot 3}{2^2} \cdot \frac{t^4}{2!} - \cdots + (-1)^n \cdot \frac{1\cdot 3\cdot 5\cdots(2n-1)}{2^n} \cdot \frac{t^{2n}}{n!} + \cdots$$

$$= \sum_{n=0}^{\infty} t^n P_n(0)$$

or $\sum_{n=0}^{\infty} (-1)^n \cdot \dfrac{1\cdot 3\cdot 5\cdots(2n-1)}{2^n} \cdot \dfrac{t^{2n}}{n!} = \sum_{n=0}^{\infty} t^n P_n(0)$

or $\sum_{n=0}^{\infty} (-1)^n \cdot \dfrac{(2n)!}{2^{2n}(n!)^2} \cdot t^{2n} = \sum_{n=0}^{\infty} t^n P_n(0)$

Equating the coefficients of t^{2n} in both sides, we get

$$P_{2n}(0) = (-1)^n \cdot \frac{(2n)!}{2^{2n}(n!)^2}.$$ □

(6) From $\sum_{n=0}^{\infty} (-1)^n \cdot \dfrac{(2n)!}{2^{2n}(n!)^2} \cdot t^{2n} = \sum_{n=0}^{\infty} t^n P_n(0)$, it is clear

that the coefficient of t^{2n+1} in the right hand side is zero, then $P_{2n+1}(0) = 0$.

(7) Replacing x by $-x$ in the generating function, we obtain

$$(1+2xt+t^2)^{-1/2} = \sum_{n=0}^{\infty} t^n P_n(-x).$$

Now, replacing t by $-t$, we get

$$(1-2xt+t^2)^{-1/2} = \sum_{n=0}^{\infty} t^n P_n(x) = \sum_{n=0}^{\infty} (-1)^n t^n P_n(-x)$$

Equating the coefficients of t^n, we obtain

$$P_n(-x) = (-1)^n P_n(x).$$ □

Example 10: Using the generating function, $(1-2xt+t^2)^{-1/2} = \sum_{n=0}^{\infty} t^n P_n(x)$, show that:

$$1 + \frac{1}{2} P_1(\cos\theta) + \frac{1}{3} P_2(\cos\theta) + \cdots = \ln\left\{\frac{1+\sin(\theta/2)}{\sin(\theta/2)}\right\}.$$

Solution: From the generating function, integrating both sides with respect to t from 0 to 1, we obtain:

$$\int_0^1 \frac{dt}{\sqrt{1-2tx+t^2}} = \sum_{n=0}^{\infty} \int_0^1 t^n P_n(x)\, dt.$$

Now, letting $x = \cos\theta$, we get

$$\int_0^1 \frac{dt}{\sqrt{1-2t\cos\theta+t^2}} = \sum_{n=0}^{\infty} P_n(\cos\theta) \int_0^1 t^n\, dt\ ^7, \text{ or}$$

$$\int_0^1 \frac{dt}{\sqrt{(t-\cos\theta)^2+\sin^2\theta}} = \sum_{n=0}^{\infty} P_n(\cos\theta) \left[\frac{t^{n+1}}{n+1}\right]_0^1\ ^8.$$

$$\sum_{n=0}^{\infty} \frac{P_n(\cos\theta)}{n+1} = \left[\ln\left\{(t-\cos\theta)+\sqrt{(t-\cos\theta)^2+\sin^2\theta}\right\}\right]_0^1$$

$$= \ln\left[(1-\cos\theta)+\sqrt{(1-\cos\theta)^2+\sin^2\theta}\right]$$

$$-\ln\left[(-\cos\theta)+\sqrt{\cos^2\theta+\sin^2\theta}\right]$$

$$= \ln\left[\frac{1-\cos\theta+\sqrt{2}\sqrt{1-\cos\theta}}{1-\cos\theta}\right]$$

$$= \ln\left[\frac{\sqrt{1-\cos\theta}+\sqrt{2}}{\sqrt{1-\cos\theta}}\right]\ ^9$$

---

7. $1-2t\cos\theta+t^2 = \sin^2\theta+\cos^2\theta-2t\cos\theta+t^2 = (t-\cos\theta)^2+\sin^2\theta$

8. $\int_0^1 \frac{dx}{\sqrt{x^2+a^2}} = \left[\ln\left\{x+\sqrt{x^2+a^2}\right\}\right]_0^1$

33

$$= \ln\left[\frac{\sqrt{2\sin^2\theta/2} + \sqrt{2}}{\sqrt{2\sin^2\theta/2}}\right] = \ln\left[\frac{1+\sin\theta/2}{\sin\theta/2}\right]. \qquad \Box$$

Example 11: From the generating function $(1-2xt+t^2)^{-1/2} = \sum_{n=0}^{\infty} t^n P_n(x)$,

show that:

$$P_n(\cos\theta) = \frac{1\cdot 3\cdot 5\cdots(2n-1)}{2^{n-1}n!}\left\{\cos n\theta + \frac{1\cdot n}{1\cdot(2n-1)}\cos[(n-2)\theta]\right.$$
$$\left.+ \frac{1\cdot 3\cdot n(n-1)}{1\cdot 2\cdot(2n-1)(2n-3)}\cos[(n-4)\theta] + \cdots\right\}.$$

Solution: From the generating function, let $x = \cos\theta = \dfrac{e^{i\theta}+e^{-i\theta}}{2}$, then

$$\left[1-t(e^{i\theta}+e^{-i\theta})+t^2\right]^{-1/2} = \left[(1-te^{i\theta})(1-te^{-i\theta})\right]^{-1/2} = \sum_{n=0}^{\infty} t^n P_n(\cos\theta)$$

Now, from the binomial expansion, we have

$$(1-te^{i\theta})^{-1/2} = 1 + \frac{1}{2}te^{i\theta} + \frac{1\cdot 3}{2\cdot 4}t^2 e^{2i\theta} + \cdots + \frac{1\cdot 3\cdots(2n-1)}{2\cdot 4\cdots(2n)}t^n e^{ni\theta} + \cdots$$

$$(1-te^{-i\theta})^{-1/2} = 1 + \frac{1}{2}te^{-i\theta} + \frac{1\cdot 3}{2\cdot 4}t^2 e^{-2i\theta} + \cdots + \frac{1\cdot 3\cdots(2n-1)}{2\cdot 4\cdots(2n)}t^n e^{-ni\theta} + \cdots$$

The coefficient of t^n in the product of these two series is

$$\frac{1\cdot 3\cdots(2n-1)}{2\cdot 4\cdots(2n)}\left(e^{ni\theta}+e^{-ni\theta}\right) + \frac{1}{2}\cdot\frac{1\cdot 3\cdots(2n-3)}{2\cdot 4\cdots(2n-2)}\left(e^{(n-2)i\theta}+e^{-(n-2)i\theta}\right)$$
$$+ \frac{1\cdot 3}{2\cdot 4}\cdot\frac{1\cdot 3\cdots(2n-5)}{2\cdot 4\cdots(2n-4)}\left(e^{(n-4)i\theta}+e^{-(n-4)i\theta}\right) + \cdots$$

Therefore,

$$P_n(\cos\theta) = \frac{1\cdot 3\cdot 5\cdots(2n-1)}{2\cdot 4\cdots(2n)}\left\{2\cos n\theta + \frac{1\cdot n}{1\cdot(2n-1)}\cdot 2\cos[(n-2)\theta]\right.$$
$$\left.+ \frac{1\cdot 3\cdot n(n-1)}{1\cdot 2\cdot(2n-1)(2n-3)}\cdot 2\cos[(n-4)\theta] + \cdots\right\}$$

or

---

19. $1 - \cos\theta = 2\sin^2\theta/2$

Legendre Polynomials and Functions

$$P_n(\cos\theta) = \frac{1\cdot 3\cdot 5\cdots(2n-1)}{2^{n-1}n!}\left\{\cos n\theta + \frac{1\cdot n}{1\cdot(2n-1)}\cos[(n-2)\theta]\right.$$
$$\left. + \frac{1\cdot 3\cdot n(n-1)}{1\cdot 2\cdot(2n-1)(2n-3)}\cos[(n-4)\theta] + \cdots\right\}$$

Note: If n is odd, the last term in the expansion of $P_n(\cos\theta)$ is a multiple of $\cos\theta$; while, if n is even, the last term is a constant. Using the expansion of $P_n(\cos\theta)$, we have

$$P_0(\cos\theta) = 1; \quad P_1(\cos\theta) = \cos\theta;$$

$$P_2(\cos\theta) = \frac{1}{4}(3\cos 2\theta + 1)$$

$$P_3(\cos\theta) = \frac{1}{8}(5\cos 3\theta + 3\cos\theta);$$

$$P_4(\cos\theta) = \frac{1}{64}(35\cos 4\theta + 20\cos 2\theta + 9)$$

$$P_5(\cos\theta) = \frac{1}{128}(63\cos 5\theta + 35\cos 3\theta + 30\cos\theta).$$

Example 12: Show that $g(x,t) = (1-2xt+t^2)^{-1/2}$ satisfies:

i. $(1-2xt+t^2)\dfrac{\partial g}{\partial t} = (x-t)g$ ii. $(1-2xt+t^2)\dfrac{\partial g}{\partial x} = tg$.

Solution: $\dfrac{\partial g}{\partial t} = \dfrac{x-t}{(1-2xt+t^2)^{3/2}}$ or $(1-2xt+t^2)\dfrac{\partial g}{\partial t} = (x-t)g$.

$\dfrac{\partial g}{\partial x} = \dfrac{t}{(1-2xt+t^2)^{3/2}}$ or $(1-2xt+t^2)\dfrac{\partial g}{\partial x} = tg$.   □

Example 13: Show that $g(x,t) = (1-2xt+t^2)^{-1/2}$ is a solution of the

differential equation $t\dfrac{\partial^2}{\partial t^2}(tg) + \dfrac{\partial}{\partial x}\left\{(1-x^2)\dfrac{\partial g}{\partial x}\right\} = 0$.

Solution: From the generating function, we have

$$g = (1-2xt+t^2)^{-1/2} = \sum_{n=0}^{\infty} t^n P_n(x), \text{ then } tg = \sum_{n=0}^{\infty} t^{n+1} P_n(x).$$

Also,

$$\dfrac{\partial g}{\partial x} = \sum_{n=0}^{\infty} t^n P_n'(x), \text{ and}$$

$$t\dfrac{\partial^2}{\partial t^2}(tg) = t\sum_{n=0}^{\infty} n(n+1)t^{n-1} P_n(x) = \sum_{n=0}^{\infty} n(n+1)t^n P_n(x)$$

$$\dfrac{\partial}{\partial x}\left\{(1-x^2)\dfrac{\partial g}{\partial x}\right\} = \dfrac{\partial}{\partial x}\left\{(1-x^2)\sum_{n=0}^{\infty} t^n P_n'(x)\right\}$$

$$= (1-x^2)\sum_{n=0}^{\infty} t^n P_n''(x) - 2x\sum_{n=0}^{\infty} t^n P_n'(x)$$

Substituting all these expressions in the given differential equation, we obtain

$$t\frac{\partial^2}{\partial t^2}(tg) + \frac{\partial}{\partial x}\left\{(1-x^2)\frac{\partial g}{\partial x}\right\}$$

$$= (1-x^2)\sum_{n=0}^{\infty} t^n P_n''(x) - 2x\sum_{n=0}^{\infty} t^n P_n'(x) + \sum_{n=0}^{\infty} n(n+1)t^n P_n(x)$$

$$= \sum_{n=0}^{\infty} t^n \left[(1-x^2)P_n''(x) - 2xP_n'(x) + n(n+1)P_n(x)\right] = 0 \qquad \square$$

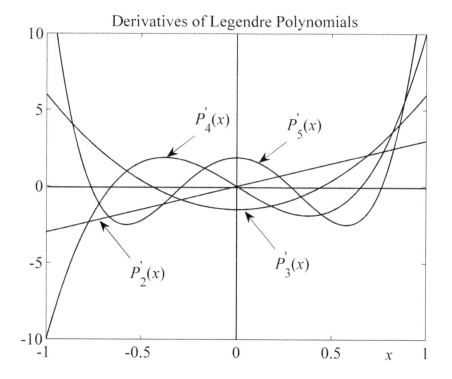

5. Recurrence Relations for Legendre Polynomials

In the previous section we have studied the generating function for Legendre Polynomials. In this section, we will derive several recurrence relations from this generating function. These recurrence relations are particularly useful for computer evaluation of Legendre Polynomials and their derivatives.

I. $\boxed{(n+1)P_{n+1}(x) - (2n+1)xP_n(x) + nP_{n-1}(x) = 0}$ \hfill (29)

Proof: From the generating function $(1 - 2xt + t^2)^{-1/2} = \sum_{n=0}^{\infty} t^n P_n(x)$,

differentiating with respect to t, we get

$$-\frac{1}{2}(1 - 2xt + t^2)^{-3/2}(-2x + 2t) = n\sum_{n=0}^{\infty} t^{n-1} P_n(x) \tag{30}$$

Multiplying both sides by $(1 - 2xt + t^2)$, we obtain

$$(x - t)(1 - 2xt + t^2)^{-1/2} = (1 - 2xt + t^2) \cdot n \sum_{n=0}^{\infty} t^{n-1} P_n(x), \text{ or}$$

$$(x - t) \sum_{n=0}^{\infty} t^n P_n(x) = (1 - 2xt + t^2) \cdot n \sum_{n=0}^{\infty} t^{n-1} P_n(x).$$

Equating the coefficients of t^n in both sides, we get

$xP_n(x) - P_{n-1}(x) = (n+1)P_{n+1}(x) - 2nxP_n(x) + (n-1)P_{n-1}(x)$ or

$(n+1)P_{n+1}(x) - (2n+1)xP_n(x) + nP_{n-1}(x) = 0$.

II. $\boxed{nP_n(x) = xP_n'(x) - P_{n-1}'(x)}$ \hfill (31)

Proof: From the generating function $(1 - 2xt + t^2)^{-1/2} = \sum_{n=0}^{\infty} t^n P_n(x)$,

differentiating with respect to x, we get

$$-\frac{1}{2}(1 - 2xt + t^2)^{-3/2}(-2t) = \sum_{n=0}^{\infty} t^n P_n'(x)$$

Legendre Polynomials and Functions

Multiplying both sides by $(2t - 2x)$, we obtain

$$-\frac{1}{2}(-2x+2t)(1-2xt+t^2)^{-3/2}(-2t) = (-2x+2t)\sum_{n=0}^{\infty} t^n P_n'(x).$$

Using equation (30) in the previous page, and rearranging, we get

$$t\sum_{n=0}^{\infty} nt^{n-1} P_n(x) = (x-t)\sum_{n=0}^{\infty} t^n P_n'(x).$$

Equating the coefficients of t^n in both sides, we get

$$nP_n(x) = xP_n'(x) - P_{n-1}'(x).$$

III. $\boxed{(2n+1)P_n(x) = P_{n+1}'(x) - P_{n-1}'(x)}$ \hfill (32)

Proof: From recurrence relation **I**, we have

$$(2n+1)xP_n(x) = (n+1)P_{n+1}(x) + nP_{n-1}(x).$$

Differentiating with respect to x, we get

$$(2n+1)xP_n'(x) + (2n+1)P_n(x) = (n+1)P_{n+1}'(x) + nP_{n-1}'(x).$$

And from recurrence relation **II**, we have

$$xP_n'(x) = nP_n(x) + P_{n-1}'(x).$$

Then from these two equations, we finally obtain

$$(2n+1)P_n(x) = P_{n+1}'(x) - P_{n-1}'(x).$$

IV. $\boxed{(n+1)P_n(x) = P_{n+1}'(x) - xP_n'(x)}$ \hfill (33)

Proof: From recurrence relations **II** and **III**, we have

$$nP_n(x) = xP_n'(x) - P_{n-1}'(x)$$

$$(2n+1)P_n(x) = P_{n+1}'(x) - P_{n-1}'(x)$$

Subtracting these two equations, we get

$$(n+1)P_n(x) = P_{n+1}'(x) - xP_n'(x).$$

V. $\boxed{(1-x^2)P_n'(x) = n\,[P_{n-1}(x) - xP_n(x)\,]}$ \hfill (34)

Proof: From recurrence relations **II** and **IV**, we have

$$nP_n(x) = xP'_n(x) - P'_{n-1}(x)$$

$$(n+1)P_n(x) = P'_{n+1}(x) - xP'_n(x)$$

Replacing n by $n-1$ in the last equation, we get

$$nP_{n-1}(x) = P'_n(x) - xP'_{n-1}(x)$$

Multiplying the first equation by x we get

$$nxP_n(x) = x^2 P'_n(x) - xP'_{n-1}(x).$$

Finally subtracting the last two equations, we obtain

$$(1-x^2)P'_n(x) = n\ [P_{n-1}(x) - xP_n(x)\ ].$$

VI. $\boxed{(1-x^2)P'_n(x) = (n+1)\ [xP_n(x) - P_{n+1}(x)\ ]}$ (35)

Proof: From recurrence relations **I** and **V**, we have

$$(2n+1)xP_n(x) = (n+1)P_{n+1}(x) + nP_{n-1}(x). \quad (i)$$

$$(1-x^2)P'_n(x) = n[P_{n-1}(x) - xP_n(x)). \quad (ii)$$

Rewriting (i) as

$$[(n+1)+n]xP_n(x) = (n+1)P_{n+1}(x) + nP_{n-1}(x)\ \text{or}$$

$$(n+1)[xP_n(x) - P_{n+1}(x)] = n[P_{n-1}(x) - xP_n(x)] \quad (iii)$$

From (ii) and (iii), we get

$$(1-x^2)P'_n(x) = (n+1)\ [xP_n(x) - P_{n+1}(x)\ ].$$

Example 14: Show that the coefficient of x^n in $P_n(x)$ is $\dfrac{1 \cdot 3 \cdot 5 \cdots (2n-1)}{n!}$.

Solution: Let c_n be the coefficient of x^n in $P_n(x)$, then from Recurrence Relation **I**, we have

$$(n+1)P_{n+1}(x) - (2n+1)xP_n(x) + nP_{n-1}(x) = 0$$

Equating the coefficient of x^{n+1} to zero, we get

$$(n+1)c_{n+1} = (2n+1)c_n\ \text{or}\ c_{n+1} = \frac{2n+1}{n+1}c_n.$$

Working backward, we obtain

$$c_n = \frac{2n-1}{n}c_{n-1} = \frac{(2n-1)(2n-3)}{n(n-1)}c_{n-2} = \cdots$$
$$= \frac{(2n-1)(2n-3)\cdots 5\cdot 3\cdot 1}{n!}c_0$$

But c_0 is the coefficient of x^0 in $P_0(x)$ which is 1.

Then $c_n = \dfrac{1\cdot 3\cdot 5\cdots (2n-1)}{n!}$. □

Example 15: If $P_n(x)$ is Legendre polynomial of degree n, and if a is such that $P_n(a) = 0$, i.e. a is a root of $P_n(x) = 0$, show that $P_{n-1}(a)$ and $P_{n+1}(a)$ are of opposite signs.

Solution: From Recurrence Relation **I**, we have
$$(n+1)P_{n+1}(x) - (2n+1)xP_n(x) + nP_{n-1}(x) = 0.$$

Letting $x = a$, then $P_n(a) = 0$ and the last equation becomes
$$(n+1)P_{n+1}(a) - (2n+1)a\cdot 0 + nP_{n-1}(a) = 0,$$
or $\dfrac{P_{n+1}(a)}{P_{n-1}(a)} = -\dfrac{n}{n+1}$.

And since n is a positive integer, then $P_{n-1}(a)$ and $P_{n+1}(a)$ are of opposite signs. □

Example 16: Show that: (Beltrami's[10] Formula):
$$(2n+1)(x^2-1)P_n'(x) = n(n+1)[P_{n+1}(x) - P_{n-1}(x)]$$

Solution: From Recurrence Relation **V**, we have
$$xP_n(x) = -\frac{(1-x^2)}{n}P_n'(x) + P_{n-1}(x).$$

Substituting for $xP_n(x)$ in Recurrence Relation **VI**, we get
$$(1-x^2)P_n'(x) = (n+1)\left[-\frac{(1-x^2)}{n}P_n'(x) + P_{n-1}(x) - P_{n+1}(x)\right].$$

Re-arranging, we obtain

---

10. Eugenio Beltrami (1835-1900)

$$(2n+1)(x^2-1)P_n'(x) = n(n+1)[P_{n+1}(x) - P_{n-1}(x)]. \qquad \square$$

Example 17: Show that: (Christoffel's[11] Expansion):

$$P_n'(x) = (2n-1)P_{n-1}(x) + (2n-5)P_{n-3}(x) + \cdots + \begin{cases} 3P_1(x) & n \text{ even} \\ P_0(x) & n \text{ odd} \end{cases}$$

Solution: In Recurrence Relation **III**, $(2n+1)P_n(x) = P_{n+1}'(x) - P_{n-1}'(x)$, replacing n by $n-1$ and re-arranging, we get

$$P_n'(x) = (2n-1)P_{n-1}(x) + P_{n-2}'(x).$$

For n even, replacing n by $n-2, n-4, \cdots, 4, 2$, we get

$$P_{n-2}'(x) = (2n-5)P_{n-3}(x) + P_{n-4}'(x)$$

$$P_{n-4}'(x) = (2n-9)P_{n-5}(x) + P_{n-6}'(x)$$

$$P_{n-6}'(x) = (2n-13)P_{n-7}(x) + P_{n-8}'(x)$$

...

$$P_4'(x) = 7P_3(x) + P_2'(x)$$

$$P_2'(x) = 3P_1(x) + P_0'(x) = 3P_1(x).$$

Adding, we get

$$P_n'(x) = (2n-1)P_{n-1}(x) + (2n-5)P_{n-3}(x) + \cdots + 7P_3(x) + 3P_1(x)$$

For n odd, replacing n by $n-2, n-4, \cdots, 5, 3$, we get

$$P_{n-2}'(x) = (2n-5)P_{n-3}(x) + P_{n-4}'(x)$$

$$P_{n-4}'(x) = (2n-9)P_{n-5}(x) + P_{n-6}'(x)$$

$$P_{n-6}'(x) = (2n-13)P_{n-7}(x) + P_{n-8}'(x)$$

...

$$P_5'(x) = 9P_4(x) + P_3'(x)$$

$$P_3'(x) = 5P_1(x) + P_1'(x) = 5P_2(x) + P_0(x).$$

Adding, we get

$$P_n'(x) = (2n-1)P_{n-1}(x) + (2n-5)P_{n-3}(x) + \cdots + 5P_2(x) + P_0(x).$$

---

11. Elwin Christoffel (1829-1900)

Legendre Polynomials and Functions

Combining the results, we finally obtain

$$P_n'(x) = (2n-1)P_{n-1}(x) + (2n-5)P_{n-3}(x) + \cdots + \begin{cases} 3P_1(x) & n \text{ even} \\ P_0(x) & n \text{ odd} \end{cases} \square$$

Example 18: Show that: (Christoffel's Summation):

$$\sum_{k=0}^{n} (2k+1)P_k(x)P_k(y) = \frac{n+1}{x-y}\left[P_{n+1}(x)P_n(y) - P_{n+1}(y)P_n(x)\right]$$

Solution: In Recurrence Relation **I**,

$(n+1)P_{n+1}(x) - (2n+1)xP_n(x) + nP_{n-1}(x) = 0$, we have

$$(2k+1)xP_k(x) = (k+1)P_{k+1}(x) - kP_{k-1}(x) \qquad (i)$$

$$(2k+1)yP_k(y) = (k+1)P_{k+1}(y) - kP_{k-1}(y) \qquad (ii)$$

Multiplying (*i*) by $P_k(y)$ and (*ii*) by $P_k(x)$ and subtracting, we get

$$(2k+1)(x-y)P_k(x)P_k(y) =$$
$$(k+1)[P_{k+1}(x)P_k(y) - P_{k+1}(y)P_k(x)]$$
$$-k[P_{k-1}(y)P_k(x) - P_{k-1}(x)P_k(y)]$$

Letting $k = 0, 1, 2, \cdots, n$ and adding, we get

$$(x-y)\sum_{k=0}^{n}(2k+1)P_k(x)P_k(y) = (n+1)\left[P_{n+1}(x)P_n(y) - P_{n+1}(y)P_n(x)\right]$$

Note that all the other terms in the right hand side will cancel out in the addition process, then

$$\sum_{k=0}^{n}(2k+1)P_k(x)P_k(y) = \frac{n+1}{x-y}\left[P_{n+1}(x)P_n(y) - P_{n+1}(y)P_n(x)\right] \square$$

Example 19: For $x > 1$, show that $P_n(x) < P_{n+1}(x)$.

Solution: First, $1 < x$ implies that $P_0(x) < P_1(x)$. We will use mathematical induction: Suppose that this result is true for n, i.e., $P_{n-1}(x) < P_n(x)$, then $\dfrac{P_{n-1}(x)}{P_n(x)} < 1$.

From Recurrence Relation **I**, we have

$(n+1)P_{n+1}(x) - (2n+1)xP_n(x) + nP_{n-1}(x) = 0$, or

$$\frac{P_{n+1}(x)}{P_n(x)} = \frac{2n+1}{n+1} \cdot x - \frac{n}{n+1} \cdot \frac{P_{n-1}(x)}{P_n(x)}$$

$$> \frac{2n+1}{n+1} \cdot x - \frac{n}{n+1} \quad \left(\text{since } \frac{P_{n-1}(x)}{P_n(x)} < 1\right)$$

$$> \frac{2n+1}{n+1} - \frac{n}{n+1} = 1 \quad (\text{since } x > 1)$$

Therefore, $P_n(x) < P_{n+1}(x)$. □

Example 20: Show that: $P'_{n+1}(x) + P'_n(x) = \sum_{k=0}^{n} (2k+1) P_k(x)$

Solution: From Recurrence Relation **III**, $(2k+1) P_k(x) = P'_{k+1}(x) - P'_{k-1}(x)$,

let $k = 1, 2, 3, \cdots, n$, we get

$3P_1(x) = P'_2(x) - P'_0(x)$, $5P_2(x) = P'_3(x) - P'_1(x)$, ...

$(2n-3) P_{n-2}(x) = P'_{n-1}(x) - P'_{n-3}(x)$,

$(2n-1) P_{n-1}(x) = P'_n(x) - P'_{n-2}(x)$,

$(2n+1) P_n(x) = P'_{n+1}(x) - P'_{n-1}(x)$.

Adding all these equations, we get

$3P_1(x) + 5P_2(x) + 7P_3(x) + \cdots + (2n+1) P_n(x)$

$= P'_{n+1}(x) + P'_n(x) - P'_0(x) - P'_1(x) = P'_{n+1}(x) + P'_n(x) - P_0(x)$

Therefore, $P'_{n+1}(x) + P'_n(x) = \sum_{k=0}^{n} (2k+1) P_k(x)$. □

Legendre Polynomials and Functions

6. Orthogonality Properties of Legendre Polynomials

Two functions $f(x)$ and $g(x)$ are said to be *orthogonal* to each other in the interval $[a, b]$ with respect to the weighting function $w(x)$ if and only if

$$\int_a^b w(x)f(x)g(x)dx = 0 \qquad (36)$$

Legendre Polynomials among many other polynomials have this orthogonality property. This is given by the following theorem.

Theorem: If m and n are non-negative integers, then

$$\int_{-1}^{1} P_m(x)P_n(x)dx = \begin{cases} 0 & \text{if } m \neq n \\ \dfrac{2}{2n+1} & \text{if } m = n \end{cases} \qquad (37)$$

The weighting function in this case is 1.

Proof: Case 1: $m \neq n$

Since $P_m(x)$ and $P_n(x)$ are solutions of Legendre Differential Equation, then omitting the argument x for convenience,

$$(1-x^2)P_m'' - 2x\,P_m' + m(m+1)P_m = 0,$$

$$(1-x^2)P_n'' - 2x\,P_n' + n(n+1)P_n = 0.$$

Multiplying the first equation by P_n and the second by P_m and subtracting, we obtain

$$(1-x^2)[P_nP_m'' - P_mP_n''] - 2x[P_nP_m' - P_mP_n']$$

$$+[m(m+1) - n(n+1)]P_nP_m = 0$$

This can be written as

$$\frac{d}{dx}\{(1-x^2)[P_nP_m' - P_mP_n']\} = [(n-m)(n+m+1)]P_nP_m.$$

Integrating both sides of this last equation with respect to x form -1 to 1, we get

$$\{(1-x^2)[P_nP_m' - P_mP_n']\}\Big|_{-1}^{1} = [(n-m)(n+m+1)]\int_{-1}^{1}P_nP_m\,dx .$$

The expression on the left hand side vanishes at both limits, and since

$m \neq n$, then $\int_{-1}^{1} P_n(x) P_m(x) dx = 0$, $m \neq n$.

Case 2: $m = n$.
Starting from the generating function, we have

$$(1 - 2xt + t^2)^{-1/2} = \sum_{n=0}^{\infty} t^n P_n(x),$$

also $(1 - 2xt + t^2)^{-1/2} = \sum_{m=0}^{\infty} t^m P_m(x)$.

Then, upon multiplication, we get

$$(1 - 2xt + t^2)^{-1} = \sum_{n=0}^{\infty} \sum_{m=0}^{\infty} t^{n+m} P_n(x) P_m(x).$$

Integration both sides with respect to x from -1 to 1, we obtain

$$\int_{-1}^{1} \frac{dx}{(1 - 2xt + t^2)} = \sum_{n=0}^{\infty} \sum_{m=0}^{\infty} \left\{ \int_{-1}^{1} P_n(x) P_m(x) dx \right\} t^{n+m},$$

And since $\int_{-1}^{1} P_n P_m dx = 0$, $m \neq n$, then the double summation to the right reduces to $\sum_{n=0}^{\infty} \left\{ \int_{-1}^{1} P_n^2(x) dx \right\} t^{2n}$.

On the other hand, we have

$$\int_{-1}^{1} \frac{dx}{(1 - 2xt + t^2)} = -\frac{1}{2t} \left[\ln(1 - 2xt + t^2) \right]_{-1}^{1} = \frac{1}{t} \left[\ln(1+t) - \ln(1-t) \right]$$

$$= \frac{1}{t} \left\{ \left(t - \frac{t^2}{2} + \frac{t^3}{3} - \cdots \right) - \left(-t - \frac{t^2}{2} - \frac{t^3}{3} - \cdots \right) \right\}$$

$$= 2 \left\{ 1 + \frac{t^2}{3} + \frac{t^4}{5} + \cdots \right\} = \sum_{n=0}^{\infty} \frac{2}{2n+1} t^{2n}.$$

Then, we have

Legendre Polynomials and Functions

$$\sum_{n=0}^{\infty}\left\{\int_{-1}^{1}P_n^2(x)dx\right\}t^{2n}=\sum_{n=0}^{\infty}\frac{2}{2n+1}t^{2n}.$$

Equating the coefficients of t^{2n} in both sides, we obtain

$$\int_{-1}^{1}P_n^2(x)dx=\frac{2}{2n+1}.$$

Note: If we let $x=\cos\theta$, the orthogonality property of Legendre Polynomials in trigonometric form will be

$$\int_{0}^{\pi}P_m(\cos\theta)P_n(\cos\theta)\sin\theta d\theta=\begin{cases}0 & \text{if } m\neq n \\ \dfrac{2}{2n+1} & \text{if } m=n\end{cases} \quad (38)$$

Example 21: Show that: $\int_{-1}^{1}x\,P_n(x)P_{n-1}(x)dx=\dfrac{2n}{4n^2-1}$.

Solution: From Recurrence Relation **I**, we have (omitting the argument for convenience) $x\,P_n=\dfrac{n+1}{2n+1}P_{n+1}+\dfrac{n}{2n+1}P_{n-1}$.

Multiplying both sides by P_{n-1} and integrating with respect to x form -1 to 1, we obtain

$$\int_{-1}^{1}x\,P_nP_{n-1}dx=\frac{n+1}{2n+1}\int_{-1}^{1}P_{n+1}P_{n-1}dx+\frac{n}{2n+1}\int_{-1}^{1}P_{n-1}^2dx.$$

From the orthogonality property of Legendre Polynomials, the first integral in the right hand side vanishes, while the second integral becomes

$$\frac{2}{2(n-1)+1}=\frac{2}{2n-1}.$$

Therofore, $\int_{-1}^{1}x\,P_n(x)P_{n-1}(x)dx=\dfrac{n}{2n+1}\cdot\dfrac{2}{2n-1}=\dfrac{2n}{4n^2-1}.$ □

Example 22: Show that

$$\int_{-1}^{1}(1-x^2)P_m'(x)P_n'(x)\,dx = \begin{cases} 0 & \text{if } m \neq n \\ \dfrac{2n(n+1)}{2n+1} & \text{if } m = n \end{cases}$$

Solution: Omitting the argument for convenience, we have

$$I = \int_{-1}^{1}(1-x^2)P_m' P_n'\,dx = \int_{-1}^{1}(1-x^2)P_n'\,dP_m.$$

Integrating by parts, we get

$$I = \int_{-1}^{1}(1-x^2)P_m' P_n'\,dx =$$

$$= \left[(1-x^2)P_n' P_m\right]_{-1}^{1} - \int_{-1}^{1} P_m\,d\left\{(1-x^2)P_n'\right\}$$

$$= -\int_{-1}^{1} P_m\left\{(1-x^2)P_n'' - 2xP_n'\right\}dx$$

The term in the square brackets vanishes at both limits, and since P_n satisfies Legendre Differential Equation, we have

$$(1-x^2)P_n'' - 2x\,P_n' = -n(n+1)P_n,$$

and the integral becomes

$$I = \int_{-1}^{1}(1-x^2)P_m' P_n'\,dx = n(n+1)\int_{-1}^{1} P_n P_m\,dx$$

Then, from the Orthogonality Property, we obtain

$$\int_{-1}^{1}(1-x^2)P_m'(x)P_n'(x)\,dx = \begin{cases} 0 & \text{if } m \neq n \\ \dfrac{2n(n+1)}{2n+1} & \text{if } m = n \end{cases} \quad \square$$

Legendre Polynomials and Functions

Example 23: Show that: $\int_0^1 x^2 P_{n+1}(x) P_{n-1}(x)\,dx = \dfrac{n(n+1)}{(2n+3)(4n^2-1)}$.

Solution: Since $x^2 P_{n+1} P_{n-1}$ is an even function, then

$$I = \int_0^1 x^2 P_{n+1} P_{n-1}\,dx = \frac{1}{2}\int_{-1}^1 x^2 P_{n+1} P_{n-1}\,dx$$

From Recurrence Relation **I**, we have

$$x P_n = \frac{n+1}{2n+1}P_{n+1} + \frac{n}{2n+1}P_{n-1}.$$

Replacing n once by $n+1$ and once by $n-1$, we obtain

$$x P_{n+1} = \frac{n+2}{2n+3}P_{n+2} + \frac{n+1}{2n+3}P_n \quad \text{and}$$

$$x P_{n-1} = \frac{n}{2n-1}P_n + \frac{n-1}{2n-1}P_{n-2}.$$

Multiplying these last two equations, we get

$$x^2 P_{n+1}P_{n-1} = \left(\frac{n+2}{2n+3}P_{n+2} + \frac{n+1}{2n+3}P_n\right)\left(\frac{n}{2n-1}P_n + \frac{n-1}{2n-1}P_{n-2}\right)$$

Integrating with respect to x from -1 to 1, and using the Orthogonality Property, we obtain

$$I = \frac{1}{2}\int_{-1}^1 x^2 P_{n+1} P_{n-1}\,dx = \frac{1}{2}\cdot\frac{n+1}{2n+3}\cdot\frac{n}{2n-1}\cdot\frac{2}{2n+1}$$

$$= \frac{n(n+1)}{(2n+3)(4n^2-1)}. \qquad \square$$

Example 24: Evaluate the following integrals:

i) $\displaystyle\int_{-1}^1 \frac{x^2\,dx}{\sqrt{5-4x}}$; ii) $\displaystyle\int_{-1}^1 \frac{(1-x^3)\,dx}{(1-x)^{3/2}}$; iii) $\displaystyle\int_{-1}^1 \frac{P_1(x)P_5(x)\,dx}{\sqrt{2-2x}}$.

Solution: From the generating function $(1-2xt+t^2)^{-1/2} = \displaystyle\sum_{n=0}^\infty t^n P_n(x)$,

i) Letting $t = 1/2$, we obtain $(5/4-x)^{-1/2} = \sum_{n=0}^{\infty} \frac{1}{2^n} P_n(x)$, or

$$(5-4x)^{-1/2} = \sum_{n=0}^{\infty} \frac{1}{2^{n+1}} P_n(x),$$

Also, we know that

$$P_2(x) = \frac{1}{2}(3x^2 - 1), \text{ then}$$

$$x^2 = \frac{1}{3} P_0(x) + \frac{2}{3} P_2(x).$$

The integral now becomes

$$\int_{-1}^{1} \frac{x^2 dx}{\sqrt{5-4x}} = \int_{-1}^{1} \left(\frac{1}{3} P_0(x) + \frac{2}{3} P_2(x) \right) \sum_{n=0}^{\infty} \frac{1}{2^{n+1}} P_n(x) dx.$$

From the orthogonality property, all terms in the integral to the right vanish except those for $n = 0$ and $n = 2$, then

$$\int_{-1}^{1} \frac{x^2 dx}{\sqrt{5-4x}} = \frac{1}{3} + \frac{1}{30} = \frac{11}{30}. \qquad \square$$

ii) $\int_{-1}^{1} \frac{(1-x^3)dx}{(1-x)^{3/2}} = \int_{-1}^{1} \frac{1+x+x^2}{\sqrt{1-x}} dx$

Letting $t = 1$, we obtain

$$\frac{1}{\sqrt{2-2x}} = \sum_{n=0}^{\infty} P_n(x), \text{ or } \frac{1}{\sqrt{1-x}} = \sqrt{2} \sum_{n=0}^{\infty} P_n(x),$$

also $1+x+x^2 = P_0 + P_1 + \frac{1}{3} P_0 + \frac{2}{3} P_2 = \frac{4}{3} P_0 + P_1 + \frac{2}{3} P_2.$

The integral now becomes,

$$\int_{-1}^{1} \frac{(1-x^3)dx}{(1-x)^{3/2}} = \int_{-1}^{1} \left(\frac{4}{3} P_0 + P_1 + \frac{2}{3} P_2 \right) \sum_{n=0}^{\infty} \sqrt{2} P_n \, dx.$$

From the orthogonality property, all terms in the integral to the right vanish except those for $n = 0$, $n = 1$ and $n = 2$, then

$$\int_{-1}^{1} \frac{(1-x^3)dx}{(1-x)^{3/2}} = \sqrt{2} \left(\frac{8}{3} + \frac{2}{3} + \frac{4}{15} \right) = \frac{18\sqrt{2}}{15}. \qquad \square$$

iii) From ii), we have

$$\frac{1}{\sqrt{2-2x}} = \sum_{n=0}^{\infty} P_n(x), \text{ also } P_1(x)P_5(x) = xP_5(x),$$

and from recurrence relation **I**, we have

$$xP_5(x) = \frac{5}{11}P_4(x) + \frac{6}{11}P_6(x),$$

then the integral becomes

$$\int_{-1}^{1} \frac{P_1(x)P_5(x)dx}{\sqrt{2-2x}} = \int_{-1}^{1} \left(\frac{5}{11}P_4(x) + \frac{6}{11}P_6(x)\right) \sum_{n=0}^{\infty} P_n(x)dx.$$

$$= \frac{10}{99} + \frac{12}{143} \approx 0.185. \quad \square$$

Example 25: If n is a positive integer, show that:

$$\int_{-1}^{1} (1-2xt+t^2)^{-1/2} P_n(x)dx = \frac{2t^n}{2n+1}.$$

Solution: From the generation function: $(1-2xt+t^2)^{-1/2} = \sum_{m=0}^{\infty} t^m P_m(x)$,

Substituting in the integral, we obtain

$$I = \int_{-1}^{1} (1-2xt+t^2)^{-1/2} P_n(x)dx = \int_{-1}^{1} P_n(x) \sum_{m=0}^{\infty} t^m P_m(x)dx.$$

From the Orthogonality Property of Legendre Polynomials, all integrals in the right hand side will vanish except when $m = n$, then

$$I = \int_{-1}^{1} (1-2xt+t^2)^{-1/2} P_n(x)dx = t^n \cdot \frac{2}{2n+1} = \frac{2t^n}{2n+1}. \quad \square$$

Example 26: Show that:

$$xP'_n(x) = nP_n(x) + (2n-3)P_{n-2}(x) + (2n-7)P_{n-4}(x) +$$
$$+ (2n-11)P_{n-6}(x) + \cdots$$

And hence deduce that: $\int_{-1}^{1} x\, P_n(x) P_n'(x)\, dx = \dfrac{2n}{2n+1}$.

Solution: From Recurrence Relation **II**: $nP_n(x) = xP_n'(x) - P_{n-1}'(x)$, we have

$$xP_n'(x) = nP_n(x) + P_{n-1}'(x) \tag{i}$$

Also, from Recurrence Relation **III**:

$$P_{n+1}'(x) = (2n+1)P_n(x) + P_{n-1}'(x),$$

Replacing n by $n-2, n-4, n-6, \cdots$, we get

$$\left.\begin{aligned} P_{n-1}'(x) &= (2n-3)P_{n-2}(x) + P_{n-3}'(x) \\ P_{n-3}'(x) &= (2n-7)P_{n-4}(x) + P_{n-5}'(x) \\ P_{n-5}'(x) &= (2n-11)P_{n-6}(x) + P_{n-7}'(x) \\ &\cdots \end{aligned}\right\} \tag{ii}$$

Adding (i) and (ii), we obtain

$$xP_n'(x) = nP_n(x) + (2n-3)P_{n-2}(x) + (2n-7)P_{n-4}(x) + \\ + (2n-11)P_{n-6}(x) + \cdots \quad \square$$

If we multiply both sides by $P_n(x)$, integrate with respect to x from -1 to 1 and use the Orthogonality Property, we get

$$\int_{-1}^{1} x\, P_n(x) P_n'(x)\, dx = n \int_{-1}^{1} x\, P_n^2(x)\, dx = \dfrac{2n}{2n+1}. \quad \square$$

Example 27: Show that: $\int_{-1}^{1} \left[P_n'(x)\right]^2 dx = n(n+1)$.

Solution: From Christoffel's Expansion:

$$P_n'(x) = (2n-1)P_{n-1}(x) + (2n-5)P_{n-3}(x) + \cdots + \begin{cases} 3P_1(x) & n \text{ even} \\ P_0(x) & n \text{ odd} \end{cases}$$

Squaring both sides, integrating with respect to x from -1 to 1 and using the Orthogonality property, we obtain

$$\int_{-1}^{1}[P'_n(x)]^2 dx = (2n-1)^2 \int_{-1}^{1} P^2_{n-1}(x)dx + (2n-5)^2 \int_{-1}^{1} P^2_{n-3}(x)dx$$

$$+ \cdots + \begin{cases} 3^2 \int_{-1}^{1} P^2_1(x)dx & n \text{ even} \\ \int_{-1}^{1} P^2_0(x)dx & n \text{ odd} \end{cases}$$

$$= \frac{2(2n-1)^2}{2(n-1)+1} + \frac{2(2n-5)^2}{2(n-3)+1} + \frac{2(2n-9)^2}{2(n-5)+1} + \cdots + \begin{cases} \frac{2 \cdot 3^2}{3} & n \text{ even} \\ 2 & n \text{ odd} \end{cases}$$

$$= 2\left[(2n-1)+(2n-5)+(2n-9)+\cdots+\begin{cases} 3 & n \text{ even} \\ 1 & n \text{ odd} \end{cases}\right]$$

This is an arithmetic progession series, then:

<u>For n even</u>: the number of terms is $n/2$, and

$$\int_{-1}^{1}[P'_n(x)]^2 dx = 2 \cdot \frac{1}{2} \cdot \frac{n}{2} \cdot [(2n-1)+3] = n(n+1).$$

<u>For n odd</u>: the number of terms is $(n+1)/2$, and

$$\int_{-1}^{1}[P'_n(x)]^2 dx = 2 \cdot \frac{1}{2} \cdot \frac{n+1}{2} \cdot [(2n-1)+1] = n(n+1).$$

Therefore, $\int_{-1}^{1}[P'_n(x)]^2 dx = n(n+1)$. □

Example 28: Show that: $\int_{-1}^{1} \frac{P_n(x)P_{n-1}(x)}{x}dx = \begin{cases} 0 & \text{if } n \text{ is even} \\ \frac{2}{n} & \text{if } n \text{ is odd} \end{cases}$.

Solution: Let: $u_n = \int_{-1}^{1} \frac{P_n(x)P_{n-1}(x)}{x}dx$, then from Recurrence Relation **I**,

$$\frac{P_n(x)}{x} = \frac{2n-1}{n}P_{n-1}(x) - \frac{n-1}{n}\frac{P_{n-2}(x)}{x}.$$

Substituting in the integral, we get,

$$u_n = \int_{-1}^{1} P_{n-1}(x)\left[\frac{2n-1}{n}P_{n-1}(x) - \frac{n-1}{n}\frac{P_{n-2}(x)}{x}\right]dx$$

$$= \frac{2n-1}{n}\cdot\frac{2}{2n-1} - \frac{n-1}{n}\int_{-1}^{1}\frac{P_{n-1}(x)P_{n-2}(x)}{x}dx$$

$$u_n = \frac{2}{n} - \frac{n-1}{n}u_{n-1}.$$

Substituting for u_{n-1}, we obtain $\boxed{u_n = \frac{n-2}{n}u_{n-2}}$.

Now, For n even: $u_n = \frac{(n-2)(n-4)\cdots 4\cdot 2}{n(n-2)(n-4)\cdots 6\cdot 4}u_2 = \frac{2}{n}u_2$, and

$$u_2 = \int_{-1}^{1}\frac{P_2(x)P_1(x)}{x}dx = \frac{1}{2}\int_{-1}^{1}\frac{x(x^2-1)}{x}dx = 0, \text{ therefore}$$

$$\int_{-1}^{1}\frac{P_n(x)P_{n-1}(x)}{x}dx = 0 \quad \text{if } n \text{ is even}.$$

For n odd: $u_n = \frac{(n-2)(n-4)\cdots 3\cdot 1}{n(n-2)(n-4)\cdots 5\cdot 3}u_1 = \frac{1}{n}u_1$, and

$$u_1 = \int_{-1}^{1}\frac{P_1(x)P_0(x)}{x}dx = \int_{-1}^{1}dx = 2,$$

therefore $\int_{-1}^{1}\frac{P_n(x)P_{n-1}(x)}{x}dx = \frac{2}{n}$ if n is odd. □

7. Integral Form of Legendre Polynomials

If n is a positive integer, Legendre polynomials are given by

$$P_n(x) = \frac{1}{\pi} \int_0^\pi \left[x \pm \sqrt{x^2 - 1} \cos\phi \right]^n d\phi \tag{39}$$

and

$$P_n(x) = \frac{1}{\pi} \int_0^\pi \frac{d\phi}{\left[x \pm \sqrt{x^2 - 1} \cos\phi \right]^{n+1}} \tag{40}$$

These two forms are known as *Laplace's first and second integral* for Legendre polynomials. To prove form (39), we proceed as follows. It may be shown by elementary methods of integral calculus that

$$\int_0^\pi \frac{d\phi}{a \pm b \cos\phi} = \frac{\pi}{\sqrt{a^2 - b^2}}, \quad a^2 > b^2 \tag{41}$$

Now, letting $a = 1 - tx$ and $b = t\sqrt{x^2 - 1}$, then

$$a^2 - b^2 = (1 - tx)^2 - t^2(x^2 - 1) = 1 - 2xt + t^2$$

Equation (41) becomes

$$\pi(1 - 2xt + t^2)^{-1/2} = \int_0^\pi \left[1 - tx \pm t\sqrt{x^2 - 1} \cos\phi \right]^{-1} d\phi$$

and from the generating function, we get

$$\pi \sum_{n=0}^\infty t^n P_n(x) = \int_0^\pi \left[1 - tx \pm t\sqrt{x^2 - 1} \cos\phi \right]^{-1} d\phi$$

Letting $z = x \pm \sqrt{x^2 - 1} \cos\phi$, then

$$\pi \sum_{n=0}^{\infty} t^n P_n(x) = \int_0^{\pi} [1-tz]^{-1} d\phi = \int_0^{\pi} \left[1 + tz + t^2 z^2 + \cdots\right] d\phi$$

$$= \int_0^{\pi} \sum_{n=0}^{\infty} [tz]^n d\phi = \sum_{n=0}^{\infty} \int_0^{\pi} \left[x \pm \sqrt{x^2-1} \cos\phi\right]^n d\phi \cdot t^n$$

Equating the coefficients of t^n in both sides, we get

$$P_n(x) = \frac{1}{\pi} \int_0^{\pi} \left[x \pm \sqrt{x^2 - 1} \cos\phi\right]^n d\phi. \qquad \square$$

To prove the Laplace's second integral for Legendre polynomials $P_n(x)$, we let $a = tx - 1$ and $b = t\sqrt{x^2 - 1}$ in equation (41), we get

$$b^2 - a^2 = (tx-1)^2 - t^2(x^2 - 1) = 1 - 2tx + t^2$$

Equation (41) becomes

$$\pi(1 - 2xt + t^2)^{-1/2} = \int_0^{\pi} \left[-1 + tx \pm t\sqrt{x^2 - 1} \cos\phi\right]^{-1} d\phi$$

or

$$\frac{\pi}{t}\left(1 - \frac{2x}{t} + \frac{1}{t^2}\right)^{-1/2} = \int_0^{\pi} \left[-1 + t\left\{x \pm \sqrt{x^2 - 1} \cos\phi\right\}\right]^{-1} d\phi \qquad (42)$$

From the generating function

Legendre Polynomials and Functions

$$(1-2xt+t^2)^{-1/2} = \sum_{n=0}^{\infty} t^n P_n(x)$$

Replacing t by $1/t$, we get

$$\left(1-\frac{2x}{t}+\frac{1}{t^2}\right)^{-1/2} = \sum_{n=0}^{\infty} \frac{1}{t^n} P_n(x)$$

Letting $z = x \pm \sqrt{x^2-1}\cos\phi$ equation (48) becomes

$$\frac{\pi}{t}\sum_{n=0}^{\infty} t^{-n} P_n(x) = \int_0^\pi [-1+tz]^{-1} d\phi = \int_0^\pi (tz)^{-1}\left[1-\frac{1}{tz}\right]^{-1} d\phi$$

$$= \int_0^\pi \frac{1}{tz}\sum_{n=0}^{\infty} [tz]^{-n} d\phi = \sum_{n=0}^{\infty} \frac{1}{t^{n+1}} \int_0^\pi \frac{d\phi}{\left[x \pm \sqrt{x^2-1}\cos\phi\right]^{n+1}}$$

Equating the coefficients of $\dfrac{1}{t^{n+1}}$ in both sides, we get

$$P_n(x) = \frac{1}{\pi}\int_0^\pi \frac{d\phi}{\left[x \pm \sqrt{x^2-1}\cos\phi\right]^{n+1}} \qquad \square$$

8. Differential Form for Legendre Polynomials (Rodrigues'[12] Formula)

One of the fundamental identities involving Legendre polynomials is Rodrigues' Formula. This formula is given in differential form

$$P_n(x) = \frac{1}{2^n n!} \cdot \frac{d^n}{dx^n}\left[(x^2-1)^n\right] \quad (43)$$

To prove this formula, we let $v = (x^2-1)^n$, then

$$\frac{dv}{dx} = 2nx(x^2-1)^{n-1}$$

Multiplying by (x^2-1), we get $(x^2-1)\dfrac{dv}{dx} = 2nx(x^2-1)^n$

or
$$(1-x^2)\frac{dv}{dx} + 2nxv = 0 \quad (44)$$

Differentiating with respect to x, we get

$$(1-x^2)v'' - 2xv' + 2nxv' + 2nv = 0$$

or
$$(1-x^2)v'' + 2x(n-1)v' + 2nv = 0$$

Differentiating n times using Leibniz's theorem, we get

$$(1-x^2)v_{n+2} - 2nxv_{n+1} - 2\cdot\frac{n(n-1)}{2!}v_n$$
$$+ 2x(n-1)v_{n+1} + 2n(n-1)v_n + 2nv_n = 0$$

or
$$(1-x^2)v_{n+2} - 2xv_{n+1} + n(n+1)v_n = 0$$

Letting $u = v_n$, we get $(1-x^2)u'' - 2xu' + n(n+1)u = 0$

Then $u = v_n$ is a solution of Legendre differential equation.

---

12. Benjamin Olinde Rodrigues (1794-1851 France)

But $u = v_n = \frac{d^n}{dx^n}\left[(x^2-1)^n\right]$. Hence u must be some multiple of $P_n(x)$, i.e.,

$$\frac{d^n}{dx^n}\left[(x^2-1)^n\right] = cP_n(x) \tag{45}$$

To determine the constant c, we know that for $x = 1$, $P_n(1) = 1$, therefore from Equation (45), we have

$$c = \frac{d^n}{dx^n}\left[(x^2-1)^n\right]\bigg|_{x=1} \tag{46}$$

But $\frac{d^n}{dx^n}\left[(x^2-1)^n\right] = \frac{d^n}{dx^n}\left[(x-1)^n \cdot (x+1)^n\right] = n!(x+1)^n + R$, where R contains $(x-1)$ as a factor, which makes $R = 0$, as $x = 1$. Then

$$c = \frac{d^n}{dx^n}\left[(x^2-1)^n\right]\bigg|_{x=1} = 2^n n!,$$

And $\quad P_n(x) = \frac{1}{2^n n!} \cdot \frac{d^n}{dx^n}\left[(x^2-1)^n\right].$ □

Rodrigues' Formula can also be derived using the summation expression for Legendre Polynomials

$$P_n(x) = \sum_{k=0}^{N} (-1)^k \cdot \frac{(2n-2k)!}{2^n k!(n-2k)!(n-k)!} x^{n-2k} \tag{47}$$

where N is the *Floor Function* $N = \begin{cases} n/2 & \text{if } n \text{ is even} \\ (n-1)/2 & \text{if } n \text{ is odd} \end{cases}$

$$P_n(x) = \frac{1}{2^n} \sum_{k=0}^{N} (-1)^k \cdot \frac{(2n-2k)(2n-2k-1)\cdots(n-2k+1)x^{n-2k}}{k!(n-k)!}.$$

Looking at the denominator in this expression, we can see that the factors in it allows us to replace the upper limit of the summation N by n, since one of the factors will vanish for every value of k greater that N. Therefore, we have

$$P_n(x) = \frac{1}{2^n} \sum_{k=0}^{n} (-1)^k \cdot \frac{(2n-2k)(2n-2k-1)\cdots(n-2k+1)x^{n-2k}}{k!(n-k)!}$$

$$= \frac{1}{2^n} \sum_{k=0}^{n} \frac{(-1)^k}{k!(n-k)!} \cdot \frac{d^n}{dx^n}\left(x^{2n-2k}\right)$$

$$= \frac{1}{2^n} \cdot \frac{d^n}{dx^n} \sum_{k=0}^{n} \frac{(-1)^k x^{2n-2k}}{k!(n-k)!}$$

$$= \frac{1}{2^n n!} \cdot \frac{d^n}{dx^n} \underbrace{\sum_{k=0}^{n} \frac{n!}{k!(n-k)!}\left(-\frac{1}{x^2}\right)^k x^{2n}}_{\left(1-\frac{1}{x^2}\right)^n \quad (binomial\ expansion)}$$

$$= \frac{1}{2^n n!} \cdot \frac{d^n}{dx^n}\left[\left(1-\frac{1}{x^2}\right)^n x^{2n}\right].$$

Hence $\quad P_n(x) = \dfrac{1}{2^n n!} \cdot \dfrac{d^n}{dx^n}\left[(x^2-1)^n\right]$.

***Example* 29:** Starting from Rodrigues' Formula, show that:

$$P_n(x) = 1 + \sum_{k=1}^{\infty} \frac{(-1)^k n(n-1)\cdots(n-k+1)\cdot(n+1)(n+2)\cdots(n+k)}{2^k (k!)^2}(1-x)^k$$

***Solution*:** From Rodrigues' Formula, we have

$$P_n(x) = \frac{1}{2^n n!} \cdot \frac{d^n}{dx^n}\left[(x^2-1)^n\right] = \frac{(-1)^n}{n!} \frac{d^n}{dx^n}\left[\frac{1}{2^n}(1-x^2)^n\right]$$

$$= \frac{(-1)^n}{n!} \frac{d^n}{dx^n}\left[(1-x)^n \cdot \frac{(1+x)^n}{2^n}\right] = \frac{(-1)^n}{n!} \frac{d^n}{dx^n}\left[(1-x)^n \cdot \left\{1-\left(\frac{1+x}{2}\right)\right\}^n\right]$$

$$= \frac{(-1)^n}{n!} \frac{d^n}{dx^n} \left[(1-x)^n \cdot \left\{ 1 - n\left(\frac{1-x}{2}\right) + \frac{n(n-1)}{2!}\left(\frac{1-x}{2}\right)^2 - \cdots \right\} \right]$$

$$= \frac{(-1)^n}{n!} \frac{d^n}{dx^n} \left[(1-x)^n - \frac{n}{2}(1-x)^{n+1} + \frac{n(n-1)}{2^2 2!}(1-x)^{n+2} - \cdots \right]$$

$$= \frac{(-1)^n}{n!} \left[(-1)^n n! - \frac{(-1)^n n}{2} \frac{(n+1)!}{1!}(1-x) + \frac{(-1)^n n(n-1)}{2^2 2!} \frac{(n+2)!}{2!}(1-x)^2 - \cdots \right]$$

$$= \left[1 - \frac{n}{2} \frac{(n+1)}{1!}(1-x) + \frac{n(n-1)}{2^2 2!} \frac{(n+1)(n+2)}{2!}(1-x)^2 - \cdots \right]$$

$$= 1 + \sum_{k=1}^{\infty} (-1)^k \frac{n(n-1)\cdots(n-k+1)\cdot(n+1)(n+2)\cdots(n+k)}{2^k (k!)} (1-x)^k \quad \square$$

Example 30: Starting from Rodrigues' Formula, prove the orthogonality property of Legendre polynomials.

Solution: From Rodrigues' Formula, $P_n(x) = \frac{1}{2^n n!} \cdot \frac{d^n}{dx^n} \left[(x^2-1)^n\right]$, we have

$$I = \int_{-1}^{1} P_m(x) P_n(x) dx$$

$$= \frac{1}{2^{n+m} n! m!} \int_{-1}^{1} \frac{d^m}{dx^m}(x^2-1)^m \frac{d^n}{dx^n}(x^2-1)^n dx$$

$$= \frac{1}{2^{n+m} n! m!} \int_{-1}^{1} \frac{d^m}{dx^m}(x^2-1)^m d\left\{\frac{d^{n-1}}{dx^{n-1}}(x^2-1)^n\right\}$$

Integrating by parts, we get,

$$I = \frac{1}{2^{n+m} n! m!} \left[\frac{d^m}{dx^m}(x^2-1)^m \cdot \frac{d^{n-1}}{dx^{n-1}}(x^2-1)^n\right]_{-1}^{1}$$

$$- \frac{1}{2^{n+m} n! m!} \int_{-1}^{1} \frac{d^{n-1}}{dx^{n-1}}(x^2-1)^n d\left\{\frac{d^m}{dx^m}(x^2-1)^m\right\}$$

The first term in this expression will vanish at both limits, then

$$I = \frac{-1}{2^{n+m} n! m!} \int_{-1}^{1} \frac{d^{n-1}}{dx^{n-1}}(x^2-1)^n \frac{d^{m+1}}{dx^{m+1}}(x^2-1)^m \, dx.$$

Now, integrating by parts $(m-1)$, we obtain

$$I = \frac{(-1)^m}{2^{n+m} n! m!} \int_{-1}^{1} \frac{d^{n-m}}{dx^{n-m}}(x^2-1)^n \frac{d^{2m}}{dx^{2m}}(x^2-1)^m \, dx.$$

But $\dfrac{d^{2m}}{dx^{2m}}(x^2-1)^m = (2m)!$, then

$$I = \frac{(-1)^m (2m)!}{2^{n+m} n! m!} \int_{-1}^{1} \frac{d^{n-m}}{dx^{n-m}}(x^2-1)^n \, dx.$$

If $m \neq n$ and $n > m$, then

$$I = \frac{(-1)^m (2m)!}{2^{n+m} n! m!} \left[\frac{d^{n-m-1}}{dx^{n-m-1}}(x^2-1)^n \right]_{-1}^{1} = 0.$$

This is the first part of the proof, for $m = n$, we have

$$I = \int_{-1}^{1} P_n^2(x) \, dx = \frac{(-1)^n (2n)!}{2^{2n} (n!)^2} \int_{-1}^{1} (x^2-1)^n \, dx.$$

The integrand of the integral to the right is an even function, so, we may write

$$I = \frac{(-1)^n (2n)!}{2^{2n-1} (n!)^2} \int_{0}^{1} (x^2-1)^n \, dx.$$

To evaluate this integral, let $x = \cos\theta$, then $dx = -\sin\theta \, d\theta$ and the limits of the integral will be from $\pi/2$ to 0, therefore

$$I = \frac{(-1)^n (2n)!}{2^{2n-1}(n!)^2} \int_0^{\pi/2} \sin^{2n}\theta \cdot \sin\theta \, d\theta$$

$$= \frac{(-1)^n (2n)!}{2^{2n-1}(n!)^2} \int_0^{\pi/2} \sin^{2n+1}\theta \, d\theta = \frac{(-1)^n (2n)!}{2^{2n-1}(n!)^2} \cdot \frac{\Gamma(n+1)\sqrt{\pi}}{2^{n+1}\Gamma(n+3/2)}$$

But $\dfrac{(2n)!}{n!} = 1 \cdot 3 \cdot 5 \cdots (2n-1)$, then

$$I = \frac{1 \cdot 3 \cdot 5 \cdots (2n-1)}{n!} \cdot \frac{2 \cdot n!\sqrt{\pi}}{(2n+1)(2n-1)\cdots 3 \cdot 1 \cdot \sqrt{\pi}} = \frac{2}{2n+1} \quad \square$$

***Example* 31:** For any n continuously differentiable function $f(x)$, show that:

$$\int_{-1}^{1} f(x) P_n(x) \, dx = \frac{(-1)^n}{2^n n!} \int_{-1}^{1} (x^2 - 1)^n f^{(n)}(x) \, dx \, .$$

***Solution*:** Substituting from Rodrigues' Formula into the integral on the right hand side, $P_n(x) = \dfrac{1}{2^n n!} \cdot \dfrac{d^n}{dx^n}\left[(x^2-1)^n\right]$, we obtain

$$I = \int_{-1}^{1} f(x) P_n(x) \, dx = \frac{1}{2^n n!} \cdot \int_{-1}^{1} f(x) \frac{d^n}{dx^n}\left[(x^2-1)^n\right] dx$$

$$= \frac{1}{2^n n!} \cdot \int_{-1}^{1} f(x) \, d\left\{\frac{d^{n-1}}{dx^{n-1}}\left[(x^2-1)^n\right]\right\}.$$

Integrating by parts, we get

$$I = \frac{1}{2^n n!} \cdot \left[f(x) \frac{d^{n-1}}{dx^{n-1}}\left[(x^2-1)^n\right]\right]_{-1}^{1}$$

$$- \frac{1}{2^n n!} \int_{-1}^{1} f'(x) \frac{d^{n-1}}{dx^{n-1}}\left[(x^2-1)^n\right] dx \, .$$

The term in the square brackets vanishes at both limits, therefore

$$I = \frac{(-1)}{2^n n!} \int_{-1}^{1} f'(x) \frac{d^{n-1}}{dx^{n-1}} \left[(x^2 - 1)^n \right] dx.$$

Integrating by parts $(n-1)$ more times, we obtain

$$I = \int_{-1}^{1} f(x) P_n(x) dx = \frac{(-1)^n}{2^n n!} \int_{-1}^{1} (x^2 - 1)^n f^{(n)}(x) dx. \qquad \square$$

Example 32: From Rodrigues' Formula, show that:

$$\int_{-1}^{1} (1 - 2xt + t^2)^{-n-1/2} (1 - x^2)^n \, dx = \frac{2^{2n+1} (n!)^2}{(2n+1)!}.$$

Solution: From **Example 25**, we have

$$\int_{-1}^{1} (1 - 2xt + t^2)^{-1/2} P_n(x) dx = \frac{2t^n}{2n+1}.$$

Plugging in the Rodrigues' Fomula for $P_n(x)$, we obtain

$$\frac{1}{2^n n!} \int_{-1}^{1} (1 - 2xt + t^2)^{-1/2} \cdot \frac{d^n}{dx^n} \left[(x^2 - 1)^n \right] dx = \frac{2t^n}{2n+1}, \text{ or}$$

$$\int_{-1}^{1} (1 - 2xt + t^2)^{-1/2} \cdot \frac{d^n}{dx^n} \left[(x^2 - 1)^n \right] dx = \frac{2^{n+1} n! t^n}{2n+1} \text{ or}$$

$$\int_{-1}^{1} (1 - 2xt + t^2)^{-1/2} d \left\{ \frac{d^{n-1}}{dx^{n-1}} \left[(x^2 - 1)^n \right] \right\} = \frac{2^{n+1} n! t^n}{2n+1}.$$

Integrating by parts, we get

$$\left[(1 - 2xt + t^2)^{-1/2} \cdot \frac{d^{n-1}}{dx^{n-1}} \left[(x^2 - 1)^n \right] \right]_{-1}^{1}$$

$$- \int_{-1}^{1} \frac{d^{n-1}}{dx^{n-1}} \left[(x^2 - 1)^n \right] d \left\{ (1 - 2xt + t^2)^{-1/2} \right\} = \frac{2^{n+1} n! t^n}{2n+1}.$$

The term in the square brackets vanishes at both end limits, then

$$-\int_{-1}^{1} \frac{d^{n-1}}{dx^{n-1}}\left[(x^2-1)^n\right]\left[t(1-2xt+t^2)^{-3/2}\right]dx = \frac{2^{n+1}n!t^n}{2n+1}, \text{ or}$$

$$-\int_{-1}^{1} \frac{d^{n-1}}{dx^{n-1}}\left[(x^2-1)^n\right]\left[(1-2xt+t^2)^{-3/2}\right]dx = \frac{2^{n+1}n!t^{n-1}}{1\cdot(2n+1)}.$$

Integrating by parts one more time, we obtain

$$(-1)^2 \int_{-1}^{1} \frac{d^{n-1}}{dx^{n-1}}\left[(x^2-1)^n\right]\left[(1-2xt+t^2)^{-5/2}\right]dx = \frac{2^{n+1}n!t^{n-2}}{1\cdot 3\cdot(2n+1)}$$

Repeating this process $n-2$ times and simplifying, we get

$$\int_{-1}^{1}(x^2-1)^n(1-2xt+t^2)^{n-1/2}\,dx = \frac{2^{n+1}n!}{1\cdot 3\cdot 5\cdots(2n-1)(2n+1)}$$

$$= \frac{2^{n+1}n!\cdot 2\cdot 4\cdot 6\cdots(2n)}{1\cdot 2\cdot 3\cdots(2n+1)} = \frac{2^{2n+1}(n!)^2}{(2n+1)!} \quad \square$$

Example 33: If $m < n$, show that:

(i) $\int_{-1}^{1} x^m P_n(x)\,dx = 0$, (ii) $\int_{-1}^{1} x^n P_n(x)\,dx = \frac{2^{n+1}(n!)^2}{(2n+1)!}$.

Solution: Substituting from Rodrigues' Formula, we get

$$I = \int_{-1}^{1} x^m P_n(x)\,dx = \frac{1}{2^n n!}\cdot \int_{-1}^{1} x^m \frac{d^n}{dx^n}\left[(x^2-1)^n\right]dx$$

$$= \frac{1}{2^n n!}\cdot \int_{-1}^{1} x^m\, d\,\frac{d^{n-1}}{dx^{n-1}}\left[(x^2-1)^n\right].$$

Integrating by parts, we get

$$I = \frac{1}{2^n n!}\cdot\left[x^m\cdot\frac{d^{n-1}}{dx^{n-1}}\left[(x^2-1)^n\right]\right]_{-1}^{1}$$

$$-\frac{m}{2^n n!}\int_{-1}^{1} x^{m-1}\frac{d^{n-1}}{dx^{n-1}}\left[(x^2-1)^n\right]dx.$$

The term in the square brackets vanishes at both limits, therefore

$$I = \frac{(-1)m}{2^n n!} \int_{-1}^{1} x^{m-1} \frac{d^{n-1}}{dx^{n-1}} \left[(x^2 - 1)^n \right] dx.$$

Integrating by parts one more times, we obtain

$$I = \frac{(-1)^2 m(m-1)}{2^n n!} \int_{-1}^{1} x^{m-2} \frac{d^{n-2}}{dx^{n-2}} \left[(x^2 - 1)^n \right] dx.$$

Integrating by parts $n-2$ more times, we obtain

$$I = \frac{(-1)^n m!}{2^n n!} \int_{-1}^{1} \frac{d^{n-m}}{dx^{n-m}} \left[(x^2 - 1)^n \right] dx$$

$$= \frac{(-1)^{n+m} m!}{2^n n!} \int_{-1}^{1} \frac{d^{n-m}}{dx^{n-m}} \left[(1-x)^n (1+x)^n \right] dx$$

(*i*) <u>For $m < n$</u>, and recalling Leibniz Theorem, we notice that the integral will vanish and

$$I = \int_{-1}^{1} x^m P_n(x) dx = 0.$$

(*ii*) <u>For $m = n$</u>, we have

$$I = \int_{-1}^{1} x^n P_n(x) dx = \frac{1}{2^n} \int_{-1}^{1} (1-x^2)^n dx = \frac{1}{2^{n-1}} \int_{0}^{1} (1-x^2)^n dx,$$

Let $x^2 = t$, $dx = \frac{dt}{2\sqrt{t}}$, and the limits for the dummy variable t will also be from 0 to 1, therefore

$$I = \frac{1}{2^{n-1}} \int_0^1 (1-x^2)^n \, dx = \frac{1}{2^{n-1}} \cdot \frac{1}{2} \int_0^1 (1-t)^n t^{-1/2} \, dx$$

$$= \frac{1}{2^n} \cdot \beta(n+1, 1/2) = \frac{1}{2^n} \cdot \frac{\Gamma(n+1)\Gamma(1/2)}{\Gamma(n+3/2)}$$

$$= \frac{1}{2^n} \cdot \frac{n!\sqrt{\pi}}{(n+1/2)(n-1/2)(n-3/2)\cdots 3/2 \cdot 1/2 \cdot \sqrt{\pi}}$$

$$= \frac{1}{2^n} \cdot \frac{2n!}{(2n+1)(2n-1)(2n-3)\cdots 3 \cdot 1} \cdot \frac{2 \cdot 4 \cdot 6 \cdots 2n}{2 \cdot 4 \cdot 6 \cdots 2n} = \frac{2^{n+1}(n!)^2}{(2n+1)!} \square$$

9. Schläfli's[13] Integral for Legendre Polynomials

From complex analysis, the nth derivative of the function $f(z)$ is given by Cauchy's Integral formula

$$\frac{d^n}{d\zeta^n} f(\zeta) = \frac{n!}{2\pi i} \int_C \frac{f(z)}{(z-\zeta)^{n+1}} dz \ . \tag{48}$$

Replacing ζ by x and z by t, we obtain

$$\frac{d^n}{dx^n} f(x) = \frac{n!}{2\pi i} \int_C \frac{f(t)}{(t-x)^{n+1}} dt \ . \tag{49}$$

Let $f(x) = (x^2 - 1)^n$, then

$$\frac{d^n}{dx^n} f(x) = \frac{d^n}{dx^n}\left[(x^2-1)^n\right] = \frac{n!}{2\pi i} \int_C \frac{(t^2-1)^n}{(t-x)^{n+1}} dt \ ,$$

From Rodrigues' Formula:

$$P_n(x) = \frac{1}{2^n n!} \cdot \frac{d^n}{dx^n}\left[(x^2-1)^n\right], \tag{50}$$

we get,

$$P_n(x) = \frac{1}{2^n n!} \cdot \frac{n!}{2\pi i} \int_C \frac{(t^2-1)^n}{(t-x)^{n+1}} dt = \frac{1}{2\pi i} \int_C \frac{(t^2-1)^n}{2^n (t-x)^{n+1}} dt$$

Legendre polynomial $P_n(x)$ are also defined by the contour integral[14]

$$P_n(x) = \frac{1}{2\pi i} \oint_C (1 - 2zx + z^2)^{-1/2} z^{-n-1} dz \ . \tag{51}$$

where C is the contour enclosing the origin and traversing in an anticlockwise direction.

---

13. Ludwig Schläfli (15 January 1814–1895 Switzerland)
14. See Arfken, G., *Mathematical Methods for Physicists, 3rd ed.*, Academic Press, Orlando, FL, 1985.

10. Associated Legendre Functions

Consider again the Legendre differential equation

$$(1-x^2)y'' - 2xy' + n(n+1)y = 0 \qquad (52)$$

$P_n(x)$ is a solution. Now, differentiating m times with respect to x using Leibniz theorem, we get

$$(1-x^2)y_{m+2} - 2xmy_{m+1} - 2\frac{m(m-1)}{2!}y_m$$

$$- 2xy_{m+1} - 2my_m + n(n+1)y_m = 0.$$

Or $(1-x^2)y_{m+2} - 2x(m+1)y_{m+1} + [n(n+1) - m(m+1)]y_m = 0$

Letting $v = y_m$, then

$$(1-x^2)v'' - 2x(m+1)v' + [n(n+1) - m(m+1)]v = 0 \qquad (53)$$

Since $P_n(x)$ is a solution of Legendre differential equation, then (53) will have a solution v given by

$$v = \frac{d^m}{dx^m} P_n(x)$$

If we let $w = v(1-x^2)^{m/2}$, then $v = w(1-x^2)^{-m/2}$.

Differentiating twice we obtain

$$v' = mxw(1-x^2)^{-1-m/2} + w'(1-x^2)^{-m/2}, \text{ and}$$

$$v'' = m(m+2)x^2 w(1-x^2)^{-2-m/2}$$

$$+ 2mxw'(1-x^2)^{-1-m/2} + w''(1-x^2)^{-m/2}$$

Substituting all these values in the differential equation (53) and simplifying, we get

$$(1-x^2)w'' - 2xw' + \left[n(n+1) - \frac{m}{1-x^2}\right]w = 0 \qquad (54)$$

Equation (54) differs from the original Legendre differential equation in that an additional term involving m appears in it. This equation is called the *Associated Legendre Differential Equation* whose solution

$$P_n^m(x) = (1-x^2)^{m/2} \frac{d^m}{dx^m}[P_n(x)], \quad m \le n \tag{55}$$

is called the *Associated Legendre Function of the First Kind*. These functions are not necessarily polynomials. But, since $P_n(x)$ is a polynomial with degree n, one can differentiate it only n times before it vanishes. This shows that $P_n^m(x)$ is defined for $m \le n$, otherwise it is zero.

It can be easily verified that

$$P_n^0(x) = P_n(x) \text{ and } P_n^m(x) = 0 \text{ if } m > n.$$

If we use Rodrigues' Formula for $P_n(x)$, and allow m to take negative values, we might write

$$P_n^m(x) = (1-x^2)^{m/2} \frac{d^m}{dx^m} \left\{ \frac{1}{2^n n!} \cdot \frac{d^n}{dx^n} \left[(x^2-1)^n \right] \right\}$$

$$= \frac{(1-x^2)^{m/2}}{2^n n!} \cdot \frac{d^{m+n}}{dx^{m+n}} \left[(x^2-1)^n \right]; \quad -n \le m \le n.$$

It can be shown that

$$P_n^{-m}(x) = (-1)^m \frac{(n-m)!}{(n+m)!} P_n^m(x). \tag{56}$$

Also, if $Q_n(x)$ is the second solution of Legendre differential equation, then

$$Q_n^m(x) = (1-x^2)^{m/2} \frac{d^m}{dx^m}[Q_n(x)]$$

will be the second solution of the associated Legendre equation. It is called the *Associated Legendre Function of the Second Kind*. The general solution of the associated Legendre differential equation can now be written as

$$y = A P_n^m(x) + B Q_n^m(x) \tag{57}$$

We state here, without proof, some of the properties of $P_n^m(x)$.

Orthogonality Property:

$$\int_{-1}^{1} P_n^m(x) P_k^m(x)\,dx = \begin{cases} 0 & \text{if } k \neq n \\ \dfrac{2(n+m)!}{(2n+1)(n-m)!} & \text{if } k = n \end{cases}$$

The angular form is given by:

$$\int_{0}^{\pi} \sin\theta \cdot P_n^m(\cos\theta) P_k^m(\cos\theta)\,dx = \begin{cases} 0 & \text{if } k \neq n \\ \dfrac{2(n+m)!}{(2n+1)(n-m)!} & \text{if } k = n \end{cases}$$

Recurrence Relations:

I. $P_n^{m+1}(x) - \dfrac{2mx}{\sqrt{1-x^2}} P_n^m(x) + [n(n+1) - m(m-1)] P_n^{m-1}(x) = 0$

II. $(2n+1)x P_n^m(x) = (n+m) P_{n-1}^m(x) + (n-m+1) P_{n+1}^m(x)$

III. $\sqrt{1-x^2}\, P_n^m(x) = \dfrac{1}{2n+1}\left[P_{n+1}^{m+1}(x) - P_{n-1}^{m+1}(x) \right]$

IV. $\sqrt{1-x^2}\, P_n^m(x) = \dfrac{1}{2n+1}\left[\begin{array}{l} (n+m)((n+m-1) P_{n-1}^{m-1}(x) \\ -(n-m+1)(n-m+2) P_{n+1}^{m-1}(x) \end{array} \right]$

Example 34: Find the general solution of

$$\dfrac{d^2 y}{d\theta^2} + \cot\theta \dfrac{dy}{d\theta} + \left[n(n+1) - \dfrac{m^2}{\sin^2\theta} \right] y = 0.$$

Solution: Let $x = \cos\theta$, then $\dfrac{dx}{d\theta} = -\sin\theta = -\sqrt{1-x^2}$,

$$\dfrac{dy}{d\theta} = \dfrac{dy}{dx} \cdot \dfrac{dx}{d\theta} = -\sqrt{1-x^2}\, \dfrac{dy}{dx}, \text{ and}$$

$$\dfrac{d^2 y}{d\theta^2} = \dfrac{d}{dx}\left[-\sqrt{1-x^2}\, \dfrac{dy}{dx} \right] \cdot \dfrac{dx}{d\theta} = (1-x^2)\dfrac{d^2 y}{dx^2} - x \dfrac{dy}{dx}$$

Substituting in the differential equation, we get

$$(1-x^2)\frac{d^2y}{dx^2} - x\frac{dy}{dx} + \frac{x}{\sqrt{1-x^2}}\left[-\sqrt{1-x^2}\frac{dy}{dx}\right]$$
$$+ \left[n(n+1) - \frac{m^2}{1-x^2}\right]y = 0$$

Or $(1-x^2)\dfrac{d^2y}{dx^2} - 2x\dfrac{dy}{dx} + \left[n(n+1) - \dfrac{m^2}{1-x^2}\right]y = 0$.

But this is the associated Legendre differential equation whose solution is $y = AP_n^m(x) + BQ_n^m(x)$. Then the general solution of the given equation is

$y = AP_n^m(\cos\theta) + BQ_n^m(\cos\theta)$. ☐

11. Series of Legendre Polynomials

If $f(x)$ and $f'(x)$ are piecewise continuous in the interval $(-1, 1)$, then there exists a Legendre series expansion of $f(x)$ of the form

$$f(x) = \sum_{n=0}^{\infty} c_n P_n(x) \tag{58}$$

To obtain the coefficients c_n, $n = 0, 1, 2, \cdots$, we multiply equation (58) by $P_m(x)$ and integrate with respect to x from -1 to 1 to obtain

$$\int_{-1}^{1} f(x) P_m(x)\, dx = \sum_{n=0}^{\infty} c_n \int_{-1}^{1} P_n(x) P_m(x)\, dx$$

and from the orthogonality property of Legendre polynomials, we get

$$\int_{-1}^{1} f(x) P_n(x)\, dx = \frac{2}{2n+1} \cdot c_n .$$

Therefore
$$c_n = \frac{2n+1}{2} \int_{-1}^{1} f(x) P_n(x)\, dx ; \quad n = 0, 1, 2, \cdots . \tag{59}$$

This representation is only valid in the interval $(-1, 1)$ since this interval is the interval of convergence of the series in equation (58).

Note: 1. If $f(x)$ is a polynomial of degree n, then

$$f(x) = \sum_{k=0}^{n} c_k P_k(x), \text{ where } c_k \text{ is given by}$$

$$c_k = \frac{2k+1}{2} \int_{-1}^{1} f(x) P_k(x)\, dx ; \quad k = 0, 1, 2, \cdots, n$$

2. If $f(x)$ is a polynomial of degree less than k, then

$$\int_{-1}^{1} f(x) P_k(x)\, dx = 0$$

3. The series $\sum_{n=0}^{\infty} c_n P_n(x)$, where

$$c_n = \frac{2n+1}{2} \int_{-1}^{1} f(x) P_n(x) \, dx$$, converges to $f(x)$ if x is not a point of discontinuity of $f(x)$ and to

$$\frac{1}{2}\left[f(x^+) + f(x^-)\right]$$ if x is a point of discontinuity.

Example 35: Find the Legendre series expansion for $f(x) = \begin{cases} -1 & -1 \leq x < 0 \\ 1 & 0 < x \leq 1 \end{cases}$.

Solution: At first glance, the function $f(x)$ is an odd function. Then we would expect that the series representation will contain only odd Legendre polynomials, then

$$c_n = \frac{2n+1}{2} \int_{-1}^{1} f(x) P_n(x) \, dx = 0, \text{ if } n = 0, 2, 4, \cdots, \text{ and}$$

$$c_n = (2n+1) \int_{0}^{1} f(x) P_n(x) \, dx, \text{ if } n = 1, 3, 5, \cdots.$$

Using the recurrence relation $P_n(x) = \frac{1}{2n+1}\left[P'_{n+1}(x) - P'_{n-1}(x)\right]$, we

get $c_n = \int_{0}^{1} \left[P'_{n+1}(x) - P'_{n-1}(x)\right] dx; \quad n = 1, 3, 5, \cdots$.

Let $n = 2k + 1$, then

$$c_{2k+1} = \int_{0}^{1} \left[P'_{2k+2}(x) - P'_{2k}(x)\right] dx; \quad k = 1, 2, 3, \cdots$$

Integrating, we get

$$c_{2k+1} = P_{2k+2}(1) - P_{2k}(1) - P_{2k+2}(0) + P_{2k}(0)$$
$$= -P_{2k+2}(0) + P_{2k}(0)$$

Now, $P_{2k}(0) = \dfrac{(-1)^k (2k)!}{2^{2k}(k!)^2}$, then

$$c_{2k+1} = \dfrac{(-1)^k (2k)!}{2^{2k}(k!)^2} - \dfrac{(-1)^{k+1}(2k+2)!}{2^{2k+2}[(k+1)!]^2}$$

$$= \dfrac{(-1)^k (2k)!}{2^{2k}(k!)^2}\left[1 + \dfrac{(2k+2)(2k+1)}{2^2(k+1)^2}\right] = \dfrac{(-1)^k (2k)!(4k+3)}{2^{2k+1}(k!)^2(k+1)}$$

Then the Legendre series expansion of $f(x)$ is

$$f(x) = \sum_{k=0}^{\infty} \dfrac{(-1)^k (2k)!(4k+3)}{2^{2k+1}(k!)^2(k+1)} P_{2k+1}(x), \quad -1 < x < 1$$

Writing the first few terms, we have

$$f(x) = \dfrac{3}{2}P_1(x) - \dfrac{7}{4}P_3(x) + \dfrac{11}{4}P_5(x) + \cdots.$$
□

Note: The first few Powers of x in terms of Legendre polynomials are

$x = P_1(x)$,

$x^2 = \dfrac{1}{3}[P_0(x) + 2P_2(x)]$,

$x^3 = \dfrac{1}{5}[3P_1(x) + 2P_3(x)]$,

$x^4 = \dfrac{1}{35}[7P_0(x) + 20P_2(x) + 8P_4(x)]$,

$x^5 = \dfrac{1}{63}[27P_1(x) + 28P_3(x) + 8P_5(x)]$,

$x^6 = \dfrac{1}{231}[33P_0(x) + 110P_2(x) + 72P_4(x) + 16P_6(x)]$.

12. Legendre Functions of the Second Kind $Q_n(x)$

Recalling the general solution of Legendre Differential equation

$$y(x) = A\left\{1 - \frac{n \cdot (n+1)}{2!}x^2 + \frac{n(n-2)\cdot(n+1)(n+3)}{4!}x^4 - \cdots\right\}$$

$$+ B\left\{x - \frac{(n-1)\cdot(n+2)}{3!}x^3 + \frac{(n-1)(n-3)\cdot(n+2)(n+4)}{5!}x^5 - \cdots\right\}$$

The *Legendre functions of the second kind* are the series solutions of Legendre differential equations that do not terminate. If n is even, then the second series does not terminate, while if n is odd the first series does not terminate.

These series solutions, apart from the multiplicative constants, define the Legendre functions of the second kind.

If we choose $B = \dfrac{(-1)^{n/2} 2^n [(n/2)!]^2}{n!}$, the Legendre function of the second kind for n even is

$$Q_n(x) = \frac{(-1)^{n/2} 2^n [(n/2)!]^2}{n!}\left\{x - \frac{(n-1)(n+2)}{3!}x^3 + \frac{(n-1)(n-3)(n+2)(n+4)}{5!}x^5 - \cdots\right\}$$

On the other hand, if we choose $A = \dfrac{(-1)^{(n+1)/2} 2^{n-1} \{[(n-1)/2]!\}^2}{1 \cdot 3 \cdot 5 \cdots n}$, the Legendre function of the second kind for n odd is

$$Q_n(x) = \frac{(-1)^{(n+1)/2} 2^{n-1} \{[(n-1)/2]!\}^2}{1 \cdot 3 \cdot 5 \cdots n}\left\{1 - \frac{n(n+1)}{2!}x^2 + \frac{n(n-2)(n+1)(n+3)}{4!}x^4 - \cdots\right\}$$

The values of A and B are chosen so that the recurrence formulas for $P_n(x)$ apply also to $Q_n(x)$.

The first few Legendre Functions of the Second Kind are

$$Q_0(x) = \frac{1}{2}\ln\left(\frac{1+x}{1-x}\right) \;;$$

$$Q_1(x) = \frac{x}{2} \ln\left(\frac{1+x}{1-x}\right) - 1;$$

$$Q_2(x) = \frac{3x^2-1}{4} \ln\left(\frac{1+x}{1-x}\right) - \frac{3x}{2};$$

$$Q_3(x) = \frac{5x^3-3x}{4} \ln\left(\frac{1+x}{1-x}\right) - \frac{5x^2}{2} + \frac{2}{3}.$$

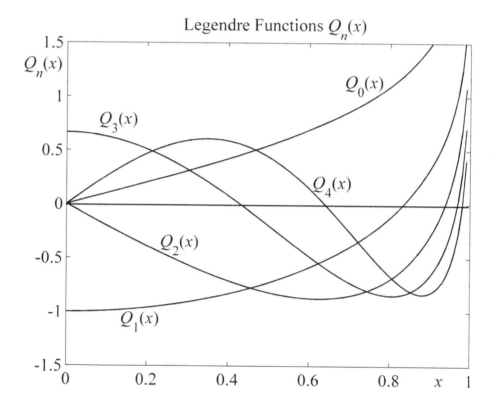

The *Associated Legendre Functions of the Second Kind* are given by

$$Q_n^m(x) = (1-x^2)^{m/2} \frac{d^m}{dx^m} Q_n(x).$$

And satisfy the Associated Legendre Differential Equation

$$(1-x^2)y'' - 2x\,y' + \left[n(n+1) - \frac{m}{1-x^2}\right] y = 0.$$

12.1. Relation between $P_n(x)$ and $Q_n(x)$

Consider the Legendre differential equation

$$(1-x^2)y'' - 2xy' + n(n+1)y = 0$$

We have established that the Legendre polynomial $P_n(x)$ is a solution. Now, to obtain the second solution $Q_n(x)$, we proceed as follows. Since $P_n(x)$ and $Q_n(x)$ are the two linearly independent solutions of the Legendre differential equation, their Wronskian should not vanish, i.e.,

$$W = \begin{vmatrix} P_n(x) & Q_n(x) \\ P_n'(x) & Q_n'(x) \end{vmatrix} \neq 0$$

then

$$W = P_n(x)Q_n'(x) - Q_n(x)P_n'(x) = P_n^2(x)\frac{d}{dx}\left(\frac{Q_n(x)}{P_n(x)}\right)$$

or

$$\frac{d}{dx}\left(\frac{Q_n(x)}{P_n(x)}\right) = \frac{W}{P_n^2(x)}$$

Integrating with respect to x from ∞ to x, we get

$$Q_n(x) = P_n(x)\int_\infty^x \frac{W}{P_n^2(x)}dx$$

We have used that fact that $\lim\limits_{x \to \infty} \frac{Q_n(x)}{P_n(x)} = 0$ (can you prove it?)

To obtain the Wronskian W, we know that $P_n(x)$ and $Q_n(x)$ satisfy the Legendre differential equation, then

$$(1-x^2)P_n''(x) - 2x P_n'(x) + n(n+1)P_n(x) = 0, \text{ and}$$

$$(1-x^2)Q_n''(x) - 2x Q_n'(x) + n(n+1)Q_n(x) = 0$$

Multiplying the first equation by $Q_n(x)$ and the second by $P_n(x)$ and subtracting, (dropping the argument for convenience) we get

$$(1-x^2)\{P_nQ_n'' - Q_nP_n''\} - 2x\{P_nQ_n' - Q_nP_n'\} = 0$$

or

$$(1-x^2)\frac{d}{dx}\{P_nQ_n' - Q_nP_n'\} - 2x\{P_nQ_n' - Q_nP_n'\} = 0$$

or

$$(1-x^2)\frac{dW}{dx} - 2xW = 0$$

This is a first order differential equation that is separable, then $W = \dfrac{c}{1-x^2}$

It can be shown that $c = 1$, then

$$Q_n(x) = P_n(x) \int_x^\infty \frac{dx}{(x^2-1)P_n^2(x)}$$

We state here a theorem that gives $Q_n(x)$ in terms of $P_n(x)$.

Theorem[15]: The Legendre function of the second kind $Q_n(x)$ is given by

$$Q_n(x) = \frac{1}{2}P_n(x)\ln\left(\frac{1+x}{1-x}\right) - \sum_{k=0}^{N}\frac{(2n-4k-1)}{(2k+1)(n-k)}P_{n-2k-1}(x),$$

where $Q_0(x) = \dfrac{1}{2}P_0(x)\ln\left(\dfrac{1+x}{1-x}\right)$, and

$$N = \begin{cases} (n-1)/2 & \text{if } n \text{ is odd} \\ (n-2)/2 & \text{if } n \text{ is even} \end{cases}$$

12.2. Properties of Legendre Functions of the Second Kind

$Q_n(x)$ satisfies the same recurrence relation as $P_n(x)$, namely

I. $\quad (n+1)Q_{n+1}(x) - (2n+1)x\,Q_n(x) + nQ_{n-1}(x) = 0$
II. $\quad nQ_n(x) = xQ_n'(x) - Q_{n-1}'(x)$
III. $\quad (2n+1)Q_n(x) = Q_{n+1}'(x) - Q_{n-1}'(x)$
IV. $\quad (n+1)Q_n(x) = Q_{n+1}'(x) - x\,Q_n'(x)$
V. $\quad (1-x^2)Q_n'(x) = n[Q_{n-1}(x) - x\,Q_n(x)]$
VI. $\quad (1-x^2)Q_n'(x) = (n+1)[x\,Q_n(x) - Q_{n+1}(x)]$

It can also be shown that:

1. *Christoffel's Second Summation Formula*:

---

15. The proof can be found in Bell, W.W., *Special Functions for Scientists and Engineers*, Van Nostrand, 1968.

$$\frac{1}{y-x} = \sum_{n=0}^{\infty}(2n+1)P_n(x)Q_n(y), \quad x>1 \text{ and } |y|\leq 1.$$

2. *Neumann's Integral Formula*:

$$Q_n(x) = \frac{1}{2}\int_{-1}^{1}\frac{P_n(x)}{x-y}dy, \quad |x|>1.$$

Example 36: Evaluate $Q_0(x), Q_1(x)$ and $Q_2(x)$.

Solution: We know that

$$Q_n(x) = P_n(x)\int_{x}^{\infty}\frac{dx}{(x^2-1)P_n^2(x)}, \text{ also}$$

$P_0(x)=1$ and $P_1(x)=x$, then

$$Q_0(x) = \int_{x}^{\infty}\frac{dx}{(x^2-1)} = \frac{1}{2}\int_{x}^{\infty}\left[\frac{1}{1+x}-\frac{1}{1-x}\right]dx = \frac{1}{2}\ln\left(\frac{1+x}{1-x}\right).$$

$$Q_1(x) = x\int_{x}^{\infty}\frac{dx}{(1-x^2)x^2} = x\int_{x}^{\infty}\left[\frac{1}{1-x^2}+\frac{1}{x^2}\right]dx = \frac{x}{2}\ln\left(\frac{1+x}{1-x}\right)-1$$

To obtain $Q_2(x)$, we use the recurrence relation **I**:

$$(n+1)Q_{n+1}(x)-(2n+1)xQ_n(x)+nQ_{n-1}(x) = 0$$

Let $n=1$, we get

$$Q_2(x) = \frac{3}{2}xQ_1(x)-\frac{1}{2}Q_0(x)$$

Substituting for $Q_1(x)$ and $Q_0(x)$, we get

$$Q_2(x) = \frac{3x^2-1}{4}\ln\left(\frac{1+x}{1-x}\right)-\frac{3x}{2}. \qquad \square$$

Example 37: Show that:
i) $$n[Q_n(x)P_{n-1}(x)-Q_{n-1}(x)P_n(x)] =$$
$$(n-1)[Q_{n-1}(x)P_{n-2}(x)-Q_{n-2}(x)P_{n-2}(x)]$$

Legendre Polynomials and Functions

ii) $P_n(x)Q_{n-1}(x) - P_{n-1}(x)Q_n(x) = \dfrac{1}{n}$

Solution: i) From the recurrence relations for $P_n(x)$ and $Q_n(x)$, we have

$$(n+1)P_{n+1}(x) - (2n+1)x\,P_n(x) + n\,P_{n-1}(x) = 0 \tag{60}$$

$$(n+1)Q_{n+1}(x) - (2n+1)x\,Q_n(x) + n\,Q_{n-1}(x) = 0 \tag{61}$$

Replacing n by $(n-1)$ in equations (60) and (61), we get

$$n\,P_n(x) - (2n-1)x\,P_{n-1}(x) + (n-1)P_{n-2}(x) = 0 \tag{62}$$

$$n\,Q_n(x) - (2n-1)x\,Q_{n-1}(x) + (n-1)Q_{n-2}(x) = 0 \tag{63}$$

Multiplying (62) by $Q_{n-1}(x)$ and (63) by $P_{n-1}(x)$, subtracting and rearranging, we get

$$n[Q_n(x)P_{n-1}(x) - Q_{n-1}(x)P_n(x)] =$$

$$(n-1)[Q_{n-1}(x)P_{n-2}(x) - Q_{n-2}(x)P_{n-2}(x)] \quad \square$$

ii) Let $u_n = n[Q_n(x)P_{n-1}(x) - Q_{n-1}(x)P_n(x)]$, then from i), we have $u_n = u_{n-1}$ and $u_n = u_{n-1} = u_{n-2} = \cdots = u_1$, then

$$n[Q_n(x)P_{n-1}(x) - Q_{n-1}(x)P_n(x)] = [Q_1(x)P_0(x) - Q_0(x)P_1(x)]$$

But $P_0(x) = 1$, $P_1(x) = x$, $Q_0(x) = \dfrac{1}{2}\ln\left(\dfrac{1+x}{1-x}\right)$ and

$$Q_1(x) = \dfrac{x}{2}\ln\left(\dfrac{1+x}{1-x}\right) - 1.$$

Substituting for all these values, we get

$$n[Q_n(x)P_{n-1}(x) - Q_{n-1}(x)P_n(x)] = -1$$

Therefore,

$$P_n(x)Q_{n-1}(x) - P_{n-1}(x)Q_n(x) = \dfrac{1}{n}. \quad \square$$

13. Shifted Legendre Polynomials

The Shifted Legendre Polynomials $\widetilde{P}_n(x)$ are orthogonal polynomials with respect to a weighing function of 1 in the interval $(0, 1)$ and

$$\int_0^1 \widetilde{P}_m(x)\widetilde{P}_n(x)\,dx = \begin{cases} 0 & \text{if } m \neq n \\ \dfrac{1}{2n+1} & \text{if } m = n \end{cases}$$

Rodrigues' Formula is given by:

$$\widetilde{P}_n(x) = \frac{1}{n!}\frac{d^n}{dx^n}\left[(x^2 - x)^n\right].$$

Also it can be shown that:

$$\widetilde{P}_n(x) = P_n(2x - 1);$$

The first few Shifted Legendre Polynomials are:

$\widetilde{P}_0(x) = 1;$ $\qquad \widetilde{P}_1(x) = 2x - 1;$

$\widetilde{P}_2(x) = 6x^2 - 6x + 1;$ $\quad \widetilde{P}_3(x) = 20x^3 - 30x^2 + 12x - 1.$

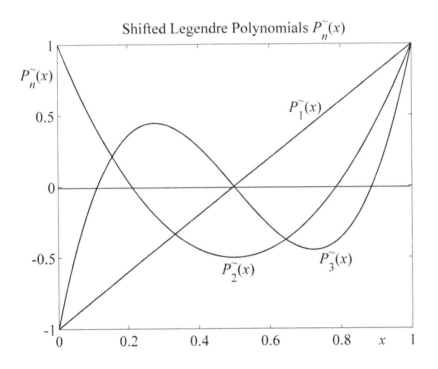

14. Summary of Legendre Polynomials and Functions

Legendre Differential equation: $(1-x^2)y'' - 2xy' + n(n+1)y = 0$

Legendre Polynomials: $P_n(x) = \sum_{k=0}^{N} (-1)^k \cdot \frac{(2n-2k)!}{2^n k!(n-2k)!(n-k)!} x^{n-2k}$

Generating Function: $(1-2xt+t^2)^{-1/2} = \sum_{n=0}^{\infty} t^n P_n(x)$

Recurrence Relations for Legendre Polynomials:

    I. $(n+1)P_{n+1}(x) - (2n+1)xP_n(x) + nP_{n-1}(x) = 0$

    II. $nP_n(x) = xP_n'(x) - P_{n-1}'(x)$

    III. $(2n+1)P_n(x) = P_{n+1}'(x) - P_{n-1}'(x)$

    IV. $(n+1)P_n(x) = P_{n+1}'(x) - xP_n'(x)$

    V. $(1-x^2)P_n'(x) = n[P_{n-1}(x) - xP_n(x)]$

    VI. $(1-x^2)P_n'(x) = (n+1)[xP_n(x) - P_{n+1}(x)]$

Orthogonal Property: $\int_{-1}^{1} P_m(x) P_n(x)\, dx = \begin{cases} 0 & \text{if } m \neq n \\ \dfrac{2}{2n+1} & \text{if } m = n \end{cases}$

$\int_{0}^{\pi} P_m(\cos\theta) P_n(\cos\theta) \sin\theta\, d\theta = \begin{cases} 0 & \text{if } m \neq n \\ \dfrac{2}{2n+1} & \text{if } m = n \end{cases}$

Integral Forms (Laplace Integrals):

$$P_n(x) = \frac{1}{\pi}\int_{0}^{\pi} \left[x \pm \sqrt{x^2-1}\cos\phi\right]^n d\phi$$

$$P_n(x) = \frac{1}{\pi}\int_{0}^{\pi} \frac{d\phi}{\left[x \pm \sqrt{x^2-1}\cos\phi\right]^{n+1}}$$

Differential Form (Rodrigues' Formula): $P_n(x) = \dfrac{1}{2^n n!} \cdot \dfrac{d^n}{dx^n}\left[(x^2-1)^n\right]$

Associated Legendre Differential Equation:

$$(1-x^2)\frac{d^2y}{dx^2} - 2x\frac{dy}{dx} + \left[n(n+1) - \frac{m^2}{1-x^2}\right]y = 0$$

Associated Legendre Functions:

$$P_n^m(x) = (1-x^2)^{m/2}\frac{d^m}{dx^m}[P_n(x)], \quad m \le n$$

$$Q_n^m(x) = (1-x^2)^{m/2}\frac{d^m}{dx^m}[Q_n(x)]$$

Recurrence Relations for the Associated Legendre Functions:

I. $P_n^{m+1}(x) - \dfrac{2mx}{\sqrt{1-x^2}} P_n^m(x) + [n(n+1) - m(m-1)]P_n^{m-1}(x) = 0$

II. $(2n+1)xP_n^m(x) = (n+m)P_{n-1}^m(x) + (n-m+1)P_{n+1}^m(x)$

III. $\sqrt{1-x^2}\, P_n^m(x) = \dfrac{1}{2n+1}\left[P_{n+1}^{m+1}(x) - P_{n-1}^{m+1}(x)\right]$

IV. $\sqrt{1-x^2}\, P_n^m(x) = \dfrac{1}{2n+1}\left[\begin{array}{l}(n+m)((n+m-1)P_{n-1}^{m-1}(x) \\ -(n-m+1)(n-m+2)P_{n+1}^{m-1}(x)\end{array}\right]$

Orthogonality Property for the Associated Legendre Functions:

$$\int_{-1}^{1} P_n^m(x) P_k^m(x)\,dx = \begin{cases} 0 & \text{if } k \ne n \\ \dfrac{2(n+m)!}{(2n+1)(n-m)!} & \text{if } k = n \end{cases}$$

$$\int_0^\pi \sin\theta \cdot P_n^m(\cos\theta) P_k^m(\cos\theta)\,dx = \begin{cases} 0 & \text{if } k \ne n \\ \dfrac{2(n+m)!}{(2n+1)(n-m)!} & \text{if } k = n \end{cases}$$

Series of Legendre Polynomials:

Legendre Polynomials and Functions

$$f(x) = \sum_{n=0}^{\infty} c_n P_n(x), \quad c_n = \frac{2n+1}{2} \int_{-1}^{1} f(x) P_n(x)\, dx; \quad n = 0, 1, 2, \cdots$$

Legendre Functions of the Second Kind:

$$Q_n(x) = P_n(x) \int_x^{\infty} \frac{dx}{(x^2 - 1) P_n^2(x)}$$

$$Q_n(x) = \frac{1}{2} P_n(x) \ln\left(\frac{1+x}{1-x}\right) - \sum_{k=0}^{N} \frac{(2n - 4k - 1)}{(2k+1)(n-k)} P_{n-2k-1}(x),$$

where $Q_0(x) = \frac{1}{2} P_0(x) \ln\left(\frac{1+x}{1-x}\right)$, and $N = \begin{cases} (n-1)/2 & \text{if } n \text{ is odd} \\ (n-2)/2 & \text{if } n \text{ is even} \end{cases}$

Recurrence Relations for Legendre Functions of the Second Kind:
  I. $(n+1) Q_{n+1}(x) - (2n+1) x\, Q_n(x) + n Q_{n-1}(x) = 0$
  II. $n Q_n(x) = x Q_n'(x) - Q_{n-1}'(x)$
  III. $(2n+1) Q_n(x) = Q_{n+1}'(x) - Q_{n-1}'(x)$
  IV. $(n+1) Q_n(x) = Q_{n+1}'(x) - x Q_n'(x)$
  V. $(1 - x^2) Q_n'(x) = n[Q_{n-1}(x) - x Q_n(x)]$
  VI. $(1 - x^2) Q_n'(x) = (n+1)[x Q_n(x) - Q_{n+1}(x)]$

Shifted Legendre Polynomials $\widetilde{P}_n(x)$

Orthogonality Property for the Shifted Legendre Functions:

$$\int_0^1 \widetilde{P}_m(x) \widetilde{P}_n(x)\, dx = \begin{cases} 0 & \text{if } m \neq n \\ \dfrac{1}{2n+1} & \text{if } m = n \end{cases}$$

Rodrigues' Formula for the Shifted Legendre Functions:

$$\widetilde{P}_n(x) = \frac{1}{n!} \frac{d^n}{dx^n}\left[(x^2 - x)^n\right].$$

Relation between $\widetilde{P}_n(x)$ and $P_n(x)$: $\widetilde{P}_n(x) = P_n(2x - 1)$

Exercises

1. If $P_0(x) = 1$ and $P_1(x) = x$, find $P_2(x)$ and $P_3(x)$.

2. Show that $\dfrac{d}{dx} P_7(x) = 13P_6(x) + 9P_4(x) + 5P_2(x) + P_0(x)$.

3. Prove that:

   a. $\displaystyle\int P_n(x)\, dx = \dfrac{1}{2n+1}\left[P_{n+1}(x) - P_{n-1}(x)\right] + c$

   b. $\displaystyle\int_x^1 P_n(x)\, dx = \dfrac{1}{n+1}\left[P_{n-1}(x) - P_{n+1}(x)\right]$

   c. $\displaystyle\sum_{k=0}^{n}(2k+1)P_k^2(x) = (n+1)^2\left[P_n(x)P_{n+1}'(x) - P_{n+1}(x)P_n'(x)\right]$
   $= (n+1)^2\left[P_n^2(x) - (x^2-1)\{P_n'(x)\}^2\right]$

4. Starting from the recurrence relation:

   $(2n+1)P_n(x) = P_{n+1}'(x) - P_{n-1}'(x)$, show that:

   $P_n'(x) = (2n-1)P_{n-1}(x) + (2n-5)P_{n-3}(x) + (2n-9)P_{n-3}(x) + \cdots$
   $= \displaystyle\sum_{k=0}^{N}(2n-4k-1)P_{n-2k-1}(x)$

   where $N = \begin{cases}(n-2)/2 & \text{if } n \text{ is even} \\ (n-1)/2 & \text{if } n \text{ is odd}\end{cases}$

5. Starting from recurrence relation:

   $(n+1)P_{n+1}(x) - (2n+1)xP_n(x) + nP_{n-1}(x) = 0$, show that:

   $P_n'(x) + P_{n-1}'(x) = \displaystyle\sum_{k=0}^{n}(2k+1)P_k(x)$

6. Using the substitution $x = \cos\theta$, show that the equation:

   $\dfrac{1}{\sin\theta}\dfrac{d}{d\theta}\left(\sin\theta\dfrac{dy}{d\theta}\right) + n(n+1)y = 0$

Legendre Polynomials and Functions

reduces to Legendre differential equation.

7. Write down the general solution of the following differential equations in terms of Legendre functions:

   a. $(1-x^2)y'' - 2xy' + 2y = 0$

   b. $(1-x^2)y'' - 2xy' + 12y = 0$

   c. $\dfrac{d^2 y}{d\theta^2} + \cot\theta \dfrac{dy}{d\theta} + 2y = 0$

8. Show that: $P_n(x) = \dfrac{1}{n!} \dfrac{\partial^n}{\partial x^n}(1 - 2xt + t^2)^{-1/2}\bigg|_{t=0}$.

9. Show that the roots of $P_n(x)$ lie between -1 and 1.

10. Show that:
$$P_n\left(-\frac{1}{2}\right) = P_0\left(-\frac{1}{2}\right)P_{2n}\left(\frac{1}{2}\right) + P_1\left(-\frac{1}{2}\right)P_{2n-1}\left(\frac{1}{2}\right) + P_2\left(-\frac{1}{2}\right)P_{2n-2}\left(\frac{1}{2}\right)$$
$$+ \cdots + P_{2n}\left(-\frac{1}{2}\right)P_0\left(\frac{1}{2}\right)$$

   Hint: Put $x = 1/2$ and $x = -1/2$ successively in the generating function, then replace t by t^2 and manipulate.

11. Show that:

   a. $P_{2n}(x) = \dfrac{(-1)^n}{2^{2n-1}} \sum\limits_{k=0}^{n} \dfrac{(-1)^k (2n+2k-1)!}{(2k)!(n+k-1)!(n-k)!} x^{2k}$

   b. $P_{2n+1}(x) = \dfrac{(-1)^n}{2^{2n}} \sum\limits_{k=0}^{n} \dfrac{(-1)^k (2n+2k+1)!}{(2k+1)!(n+k)!(n-k)!} x^{2k+1}$

12. Show that:

   a. $\displaystyle\int_0^\pi P_n(\cos\theta) \cos n\theta\, d\theta = \beta\left(n + \frac{1}{2}, \frac{1}{2}\right)$ if n is a positive integer.

b. $\int_0^\pi P_n(\cos\theta)\cos n\theta \, d\theta = \dfrac{1\cdot 3\cdot 5 \cdots (2n-1)}{2\cdot 4\cdot 6 \cdots 2n}$.

c. $\sum_{n=0}^{\infty} P_n(\cos\theta) = \operatorname{cosec}\dfrac{\theta}{2}$.

13. Show that $|P_n(\cos\theta)| \leq 1$ when θ is real.

14. Show that:

a. $\int_{-1}^{1} x\, P_n(x)\, dx = \begin{cases} 0 & \text{if } n \neq 1 \\ 2/3 & \text{if } n = 1 \end{cases}$

b. $\int_{-1}^{1} P_n(x) P'_{n+1}(x)\, dx = 2, \quad n = 0, 1, 2, 3, \cdots$

c. $\int_{-1}^{1} xP_n(x) P'_{n+1}(x)\, dx = \dfrac{2n}{2n+1}, \quad n = 0, 1, 2, 3, \cdots$

d. $\int_{-1}^{1} (1-x^2) P'_n(x) P'_m(x)\, dx = 0, \quad m \neq n$

e. $\int_{-1}^{1} x^2 P_{n+1}(x) P_{n-1}(x)\, dx = \dfrac{2n(n+1)}{(2n-1)(2n+1)(2n+3)}$.

15. Show that $\dfrac{1-t^2}{(1-2xt+t^2)^{3/2}} = \sum_{n=0}^{\infty} (2n+1) t^n P_n(x)$.

16. Show that: $\int_0^1 P_{2n}(x) P_{2n+1}(x)\, dx = \int_0^1 P_{2n}(x) P_{2n-1}(x)\, dx$.

17. Use recurrence relation **V** and **VI** to show that:

$(x^2 - 1) P'_n(x) = \dfrac{n(n+1)}{2n+1}[P_{n+1}(x) - P_{n-1}(x)]$,

Then deduce that $\int_{-1}^{1} (x^2-1)P_{n+1}(x)P_n'(x)\,dx = \dfrac{2n(n+1)}{(2n+1)(2n+3)}$.

18. Show that: $\sin^n \theta P_n(\sin\theta) = \sum_{k=0}^{n} \dfrac{(-1)^k n!}{m!(n-m)!}\cos^m \theta P_m(\cos\theta)$.

19. Show that:

$(n+1)[P_n(x)P_{n+1}'(x) - P_{n+1}(x)P_n'(x)] = (n+1)^2 P_n^2(x) - (x^2-1)P_n'^2(x)$

20. Show that:

$P_{2n+1}'(x) = (2n+1)P_{2n}(x) + 2nx P_{2n-1}(x) + (2n-1)x^2 P_{2n-2}(x)$

$+ \cdots + 2x^{2n-1} P_1(x) + x^{2n}$

21. Show that: $\sum_{k=0}^{\infty} \dfrac{x^{k+1} P_k(x)}{k+1} = \dfrac{1}{2}\ln\left(\dfrac{1+x}{1-x}\right)$.

22. Show that $\dfrac{1+t}{t(1-2xt+t^2)^{1/2}} - \dfrac{1}{t} = \sum_{n=0}^{\infty} t^n [P_n(x) + P_{n+1}(x)]$.

23. If $-1 < x < 1$ and n is a positive integer, show that:

a. $|P_n(x)| < 1$ b. $|P_n(x)| < \sqrt{\dfrac{\pi}{2n(1-x^2)}}$

Hint: Use Laplace first integral.

24. Using Rodrigue's formula, show that:

$P_{n+1}'(x) - P_{n-1}'(x) = (2n+1)P_n(x)$ and hence deduce that:

$\int_x^1 P_n(x)\,dx = \dfrac{1}{2n+1}[P_{n-1}(x) - P_{n+1}(x)]$.

25. Show that $P_n(x)Q_{n-2}(x) - P_{n-2}(x)Q_n(x) = \dfrac{2n-1}{n(n-1)}$.

26. Expand $x^4 - 3x^2 + x$ in a series of Legendre polynomials.

27. Expand $f(x) = \begin{cases} 2x+1 & 0 < x \leq 1 \\ 0 & -1 \leq x < 0 \end{cases}$ in a series of Legendre polynomials.

28. Expand $f(x) = \begin{cases} 1 & 0 < x \leq 1 \\ 0 & -1 \leq x < 0 \end{cases}$ in a series of Legendre polynomials.

29. Expand $f(x) = x^2$ in a series of Legendre polynomials.

30. If $f(x) = \sum_{k=0}^{\infty} c_k P_k(x)$, show that $\int_{-1}^{1} [f(x)]^2 dx = \sum_{k=0}^{\infty} \frac{c_k^2}{2k+1}$.

31. Show that:

a. $x^{2m} = \sum_{n=0}^{m} \frac{2^{2n}(4n+1)(2m)!(m+n)!}{(2m+2n+1)!(m-n)!} P_{2n}(x)$

b. $x^{2m+1} = \sum_{n=0}^{m} \frac{2^{2n+1}(4n+3)(2m+1)!(m+n+1)!}{(2m+2n+3)!(m-n)!} P_{2n+1}(x)$

32. Find the Legendre series expansion for $f(x) = \cos^2 x$, $0 \leq x \leq \pi$.

33. Using Rodrigues' Formula, show that:
$(2n+1)P_n(x) = P'_{n+1}(x) - P'_{n-1}(x)$.

34. Prove Christoffel's Second Summation Formula:

$$\frac{1}{y-x} = \sum_{n=0}^{\infty} (2n+1)P_n(x)Q_n(y)$$

35. Prove Neumman's Integral Formula: $Q_n(x) = \frac{1}{2} \int_{-1}^{1} \frac{P_n(x)}{y-x} dx$, $y > 1$.

36. From the summation expression for $Q_n(x)$, show that:

$\frac{d^{n+1}}{dx^{n+1}} Q_n(x) = \frac{(-1)^n 2^n n!}{(x^2-1)^{n+1}}$. **Hint:** Use Laplace first integral.

37. Show that: $Q_n(x) = \int_0^{\infty} \frac{dt}{\left[x + \Gamma(x^2-1)\cosh t\right]^{n+1}}$.

References

[1] Abramowitz, M. and Stegun, I.A. (Eds.), *Handbook of Mathematical Functions with Formulas, Graphs, and Mathematical Tables, 9th printing*, Dover Publications, New York, 1972, also John Wiley & Sons Inc; Reprint edition, **ISBN**: 0471800074, September 1993, 1060 pages.

[2] Andrews, George E., Askey, Richard and Roy, Ranjan, *Special Functions*, Cambridge University Press, 1st ed., **ISBN**: 0521789885, February 2001, 620 pages. Also **ISBN**: 0521623219.

[3] Andrews, George E., *Special Functions*, Cambridge University Press, ISBN: 0521623219, January 1999, 664 pages.

[4] Andrews, Larry C., *Special Functions of Mathematics for Engineers*, Second Edition, SPIE-International Society for Optical Engine, November, 1997, ISBN: 0819426164, 504 pages.

[5] Andrews, Larry C., *Special Functions for Engineers and Applied Mathematics*, New York, Macmillan, ISBN: 0029486505, January 1985.

[6] Andrews, Larry C., *Special Functions of Mathematics for Engineers*, McGraw-Hill, ISBN: 0070018480, October 1991.

[7] Arfken, George B. and Weber, Hans, *Mathematical Methods for Physicists*, Academic Press, 5th edition, October, 2000, ISBN: 0120598256, 1112 pages.

[8] Askey, Richard A. (Editor), *Theory and Application of Special Functions*, Proceedings of an Advanced Seminar Sponsored by the Mathematics Research Center, the University of Wisconsin-Madison, March 21-April 2, 1975, New York, Academic Press, ISNB: 0120648504, June 1975, 560 pages.

[9] Askey, Richard A., *Orthogonal Polynomials and Special Functions*, Regional Conference Series in Applied Mathematics 21, SIAM, Philadelphia, ISBN: 0898710189, June 1975, 110 pages.

[10] Askey, Richard A., Schempp, W. and Koornwinder, T.H, *Special Functions: Group Theoretical Aspects and Applications*, (Mathematics and Its Applications), Kluwer Academic Publishers, January, 2002, **ISBN**: 1402003196, 348 pages.

[11] Bailey, W.N., "On the Product of Two Legendre Polynomials." *Proc. Cambridge Philos. Soc.* **29**, 173-177, 1933.

[12] Barnerji, P.K. (Editor), *Special Functions: Selected Articles*, ISBN: 817233267X.

[13] Bell, William Wallace, *Special Functions for Scientists and Engineers*,

London, van Nostrand Comp. Ltd., ISBN: 0442006829, 1968. 247 pages; also Dover Publications, July, 2004, ISBN: 0486435210, 272 pages.

[14] Belousov, S. L., *Tables of normalized Associated Legendre Polynomials*, New York, MacMillan Company, 1962, 379 pages.

[15] Beyer, W.H., *CRC Standard Mathematical Tables, 28th ed*, CRC Press, Boca Raton, FL, 1987.

[16] Boas, Mary L., *Mathematical Methods in the Physical Sciences*, John Wiley & Sons; 3 edition, ISBN: 0471198269, 2005, 864 pages.

[17] Briggs, Lyman J. and Lowan, Arnold N., *Tables of Associated Legendre Functions*, Columbia University Press, New York, NY, 1945, 303 pages.

[18] Brychkov, Yu.A. and Prudnikov, P.A., *Integrals and Series: More Special Functions*, T&F STM; 3rd edition, ISBN: 2881246826, January 1990, 800 pages.

[19] Bustoz, Joaquin, Ismail, Mourad E.H. and Suslov, S.K. (Editors), *Special Functions 2000: Current Perspective and Future Directions*, Proceedings of the NATO Advanced Study Institute on Special Functions 2000, held in Tempe, Arizona, USA, NATO Science Series II: Mathematics, Physics and Chemistry , Vol. 30, Kluwer Academic Pub., ISBN: 0-7923-7120-8, August, 2001, 536 pages.

[20] Carlson, Bille Chandler, *Special Functions of Applied Mathematics*, New York, Academic Press, ISBN: 0121601501, June 1977, 335 pages.

[21] Chakrabarti, A., *Elements of Ordinary Differential Equations and Special Functions*, Wiley, ISBN: 0470216409, January 1990, 148 Pages, also New Age International (P) Ltd. , New Delhi, ISBN: 812240880X, 2002, 158 pages.

[22] Chihara, Theodore Seio, *An Introduction to Orthogonal Polynomials*, Gordon and Breach Publications, New York, **ISBN:** 0677041500, April 1978, 250 pages.

[23] Dieudonne, Jean A., Special Functions and Linear Representations of Lie Groups, Amer Mathematical Society, ISBN: 0821816926, May 1980, 59 pages.

[24] Erdélyi, Arthur, Magnus, W., Oberhettinger, F. and Tricomi, F.G., *Higher Transcendental Fuctions* vols. 1-3, McGraw-Hill, New York, 1953, also 5 vols., Krieger Publishing, Melbourne, Fla., ISBN: 0898742072, June 1981.

[25] Gormley, P.G., *The Zeros of Legendre Functions,* Proceedings of the Royal Irish Academy, Vol. XLIV, Sec. A, no. 4, Dublin 1937.

[26] Gumprecht, R.O. and Sliepcevich, C.M., *Functions of Partial Derivatives of Legendre Polynomials*, University of Michigan Press, Ann Arbor, MI, 1951, 310 pages.

[27] Gupta, B.D., *Mathematical Methods with Special Functions*, Stosius Inc/Advent Books Division, ISBN: 8122002684, 1992, 906 pages.

[28] Henrici, Peter, *Applied and Computational Complex Analysis: Special Functions, Integral Transforms, Asymptotics, Continued Fractions*, John Wiley & Sons Inc., ISBN: 047154289X, March 1991, 672 pages.

[29] Hochstadt, Harry, *Special Functions of Mathematical Physics*, Dover Publications, New York, ISBN: 0486652149, 1986, 332 pages.

[30] Jahnke, E., Emde, F. and Lösch, F. *Tables of Higher Functions*, McGraw-Hill, New York, 1960.

[31] Johnson D.E. and Johnson, Johnny R., *Mathematical Problems in Engineering & Physics: Special Functions & Boundary Value Problems*, Ronald P., US, ISBN: 0826047904, December 1965.

[32] Lagrange, René. *Polynomes et fonctions de Legendre*. Paris, Gauthier-Villars, 1939.

[33] Lebedev, Nikolai Nikolaevich, *Special Functions and Their Applications*, Prentice-Hall, Englewood Cliffs, New Jersey, **ASIN:** B0006BMX4E 1965, 308 pages; also revised English edition, New York, Dover Publications, ISBN: 0486606244, 1972, 308 pages.

[34] Legendre, A.M. "Sur l'attraction des Sphéroides." *Mém. Math. et Phys. présentés à l'Ac. r. des. sc. par divers savants* **10**, 1785.

[35] Levin, A. and Lubinsky, D.S., "Christoffel Functions and Orthogonal Polynomials for Exponential Weights on [-1,1]", *Memoirs of the American Mathematical Society*, No.535, Vol. 11, 1994.

[36] Luke, Yudell L., *The Special Functions and their Approximations*, Vols. 1 and 2, New York, Academic Press, 1969.

[37] MacRobert, Thomas Murray, *Spherical Harmonics*, Pergamon Press; 3rd edition, 1967, 349 pages.

[38] Magnus, W.; Oberhettinger, F.; and Soni, R.P., *Formulas and Theorems for the Special Functions of Mathematical Physics*, 3rd ed., Springer-Verlag, New York, ISBN: 0387035184, 1966, 508 pages.

[39] H.L Manocha, H.L.; Srinivasa Rao, K.; and Agarwal, R.P., *Selected Topics in Special Functions*, Allied, New Delhi, ISBN: 8177641697.

[40] Marcellan, Francisco, *Laredo Lectures on Orthogonal Polynomials and Special Functions*, Nova Science Pub. Inc., ISBN: 1594540098, June 2004, 210 pages.

[41] Margenau, Henry and Murphy, G.M., *The Mathematics of Physics and Chemistry*, R. E. Krieger Pub. Co; 2d ed edition (1976), **ISBN:** 0882754238, 604 pages.

[42] National Bureau of Standards, *Tables of Associated Legendre functions*, Columbia University Press 1945, 354 pages.

[43] Nikiforov, A.F. and Uvarov, V.B., *Special Functions of Mathematical Physics*, Translated from the Russian by R. P. Boas, Birkhauser, Basel, ISBN: 3764331836, 1988.

[44] Nikiforov, Arnold F.; and Uvarov, V.B., *Special Functions of Mathematical Physics: A Unified Introduction With Applications*, **ISBN:** 0817631836, Birkhauser Boston, February, 1988, 448 pages.

[45] Olver, Frank W.J., *Asymptotics and Special Functions*, AK Peters Ltd, 2nd edition, **ISBN:** 1568810695, June 1997, also Academic Press, ISBN: 012525850X, 1974.

[46] Press, W.H., Flannery, B.P., Teukolsky, S.A., and Vetterling, W.T., *Numerical Recipes in FORTRAN: The Art of Scientific Computing, 2nd ed.*, Cambridge, England, Cambridge University Press, 1992.

[47] Rainville, Earl David, *Special Functions*, Chelsea, New York, **ISBN:** 0828402582, June 1971, 365 pages.

[48] Rao Srinivasa K. et al, *Special Functions and Differential Equations*, Proceedings of a Workshop Held at the Institute of Mathematical Sciences, Chennai, India During 13-24 January 1997, Allied, New Delhi, ISBN: 8170237645, 1997.

[49] Ross, Bertram and Farrell, Orin J., *Solved Problems in Analysis: As Applied to Gamma, Beta, Legendre & Bessel Function*, Peter Smith Publisher Inc., ISBN: 0844600911, January 1984; also Dover Publications, ISBN: 0486627136, June, 1974, 410 pages.

[50] Sansone, Giovanni, *Orthogonal Functions,* rev. English ed., New York, Dover, 0486667308, June 1991, 411 pages.

[51] Siegel, K.M.; Brown, D.M.; H. E. Hunter, H.E. et al., *The Zeros of the associated legendre functions* $P_n^m(x)$ *of non- integral degree*, Ann Arbor, MI Willow Run Research Center, Engineering Research Institute, 1953, 30 pages.

[52] Singh P. and Denis, R.Y., *Special Functions and their Applications*, Dominant Publishers and Distributors, New Delhi, ISBN: 8187336919, 2001, 192 pages.

[53] Sneddon, Ian Naismith, *Special Functions of Mathematical Physics and Chemistry*, 2nd ed., Oliver and Boyd, Edinburgh, **ISBN:** 0582443962,

1961, also Oliver and Boyd, 1956; also Longman Group United Kingdom, ISBN: 0582443962, 1980.

[54] Snow, Chester, *Hypergeometric and Legendre Functions with Applications to Integral Equations of Potential Theory*. Washington, DC. US Government Printing Office, 1952, 319 Pages.

[55] Sternberg, Wolfgang and Smith, Turner Linn, *The Theory of Potential and Spherical Harmonics*, 2nd ed., Toronto, University of Toronto Press, 1944, 312 pages.

[56] Talman, James D., *Special Functions: A Group Theoretic Approach*, New York, W.A. Benjamin, 1968, 260 pages.

[57] Temme, Nico M., *Special Functions: An Introduction to the Classical Functions of Mathematical Physics*, New York, John Wiley & Sons Inc., ISBN: 0471113131, 1996. 374 pages.

[58] Truesdell, Clifford, *An Essay Toward a Unified Theory of Special Functions: Based upon the Functional Equation*, Princeton Univ Pr, ISBN: 0691095779, 1948, 182 pages.

[59] Virchenko, Nina O., *Generalized Associated Legendre Functions and Their Applications,* World Scientific Pub. Co. Inc., ISBN: 9810243537, 2001.

[60] Wang, Z.X., and Guo, D.R., *Special Functions*, Singapore, World Scientific Pub. Co., ISBN: 997150667X, October 1989. 710 pages.

[61] Whittaker, E.T. and Watson, G.N., *A Course in Modern Analysis, 4th ed.*, Cambridge University Press, Cambridge, England, 1990.

[62] Zhang, Shanjie and Jin, Jianming, *Computation of Special Functions*, New York, Wiley-Interscience; ISBN: 0471119636, July, 1996. 717p.

[63] Zhurina, M. I. and Karmazina, L. N, *Tables of the Legendre Functions: Parts 1 and 2*, New York, The Macmillan Company, 1964.

CPSIA information can be obtained
at www.ICGtesting.com
Printed in the USA
BVOW07s2003170316
440776BV00008B/72/P